From Atom to Kosmos

Cover design by MADHU

From Atom to Kosmos

Journey Without End

L. GORDON PLUMMER

This publication made possible with the assistance of the Kern Foundation

A co-publication of
Point Loma Publications, Inc.
San Diego, California
and
The Theosophical Publishing House
Wheaton, Ill. U.S.A.
Madras, India — London, England

First published 1940. Second Revised
Edition 1987
First Quest Edition 1989, co-published by Point Loma Publications, San Diego, California and The Theosophical Publishing House, Wheaton, Illinois, a department of the Theosophical Society in America.

Library of Congress Cataloging-in-Publication Data

Plummer, L. Gordon.
From atom to kosmos.
(A Quest book)
Includes index.
1. Theosophy. I. Title.
BP565.P4F76 1989 299'.934 88-29604
ISBN 0-8356-0308-3

Printed in the United States of America

Dedicated to
G. de P.
without the inspiration of whose teachings
this book would never have been written

Contents

Preface

The reader of this book is entitled to a word of explanation. In the first place, it is written for all, for student and layman alike. It is not intended as an exhaustive treatise, but we hope it may stand as a pointer that will lead the serious reader to a further study of that age-old wisdom today called Theosophy.

It has been the attempt to present the fundamentals of theosophical teaching as the writer has understood them. Each student approaches his subject in his own way, and it is our belief that the more copious the literature — which means the greater the variety in manner of presentation, *provided that strict adherence to the original teachings is maintained* — the greater is the opportunity for the realization of our ideal, which is that Theosophy shall in good time become universally known and loved.

The first edition of this present study appeared some decades ago. Because of its warm reception then, particularly as a text in study groups in the USA and abroad, it is hoped that the second edition put out by Point Loma Publications and this new Quest edition will be welcomed with equal understanding, for it is probable that at no time in recorded history has there been greater opportunity than today for the promulgation of the doctrines of the esoteric philosophy. The new editions also make possible the incorporation of a few minor corrections in view of the advances particularly in the field of science.

It is, of course, impossible to do more in a work of this kind than touch upon certain salient aspects of the ancient Wisdom-Religion, and the reader is urged to pursue further studies with the help of the standard theosophical works of Helena Petrovna Blavatsky, William Q. Judge, and G. de Purucker.

May this study be received in the same spirit of devotion to truth in which it is offered.

I

By Way of Introduction

Have you ever looked through a telescope at the stars and been spellbound at the wonders of the heavens? Have you not considered Sirius, Orion, and the Pleiades? Perchance you wonder how these things came into being, and would like to take the means at hand to know the marvels of nature.

Seeing is believing, and to the end that we may see, we shall don the winged cap and sandals of Hermes, and shall soar far above the Earth, beyond the pale moon, passing at a step the confines of the solar system, until we find ourselves lost in the velvet blackness of the night sky. Far, far below, we see the sun, a dazzling globe of light. We can discover only the outermost of the planets, those moving in smaller orbits having been lost against the brilliance of the sun. As we speed on toward a distant goal, the last of the planets disappear, and the sun has dwindled down to the size of a pinpoint. And now we seem to be at the center of a vast celestial sphere, with all the stars at equal distances from us. This being only an appearance, it rapidly changes. For as we move with the speed of thought, we find that one star out of those myriads grows brighter. We are passing close to another solar system! Perhaps we pass right through it, seeing new planets circling about in new orbits, but we do not pause in our flight. Were we traveling at the speed of light, it might be years before we pass through another such solar system. But we can think ourselves at will on the outskirts of the Milky Way, and our own pinpoint sun has long since been lost among the stars of some far-off constellation. There are no definite boundaries to the Milky Way, yet in time we find ourselves with our own great galactic system at our backs, and well-nigh endless space before us, dotted here and there by other galaxies.

Let us not pause, however, for our one thought is to put distance between our own universe and ourselves, and so in a trice our whole galaxy has become but a faint nebulous gleam in a constellation which is part of some faraway galactic system. And yet we speed on. Glancing back, we see that our whole Milky Way is but a cloud of light; and then it disappears altogether, and all about us are stars, stars, stars.

Where has our journey led us? What new knowledge have we grasped? None other than the fact that there can be no end to the spaces of Space. Never can we reach a point in the heavens of which we may say: "There can be no more stars." Journeying home once more to our home planet, we ask on our way: "Whence came all these stars? Was there ever a time when they were not? What are the laws whereby they move? If laws imply a lawgiver, who or what is their lawgiver? Where is he, that he can hold sway over that which is admittedly infinite? What is life, and how did it come into being? Are life and death separate, or are the two eternally linked, making all things alive? We know that *we* are alive, but if the universe is also alive, what differences are there between it and ourselves, other than size? Do we belong to nature only because we are on the earth, or did we belong to it before birth? Are we eternally a part of nature, and if so, why do we die, and what are the mysteries of death?"

We find that in our journey we have not touched these deeper issues at all. Is then, this boundless universe over which we have traveled in thought so limited that it can give no answer to these questions? There is an intuition that tells us that answers do exist, else the human mind would not be inspired to ask questions. Yes, we must make another journey, this time into the invisible realms of *thought;* and with Theosophy as the key, and the pages which follow as our guide, let us now contemplate new vistas, and learn what may be known of the truths concerning the universe.

II

Whence Came the Stars?

Like travelers at the crossroads we stand in momentary indecision; for before us lie two courses. In the one, we pursue intellectually a knowledge of the operation of certain laws, which bring about the interplay of forces, which in their turn result in the physical phenomena comprising the visible universe; and in the other method of consideration we see Intelligence, a small portion of whose nature appears to us as the so-called laws of nature, at work behind the scenes. We will choose the first of these courses, and if it leads to a satisfactory explanation of the origin and destiny of the universe, we can ask for no more. But if it leaves questions unanswered, we must retrace our steps and seek the other pathway, for one or the other will lead us to our goal, which is true knowledge about nature.

We see about us, then, a universe in which the law of mutual attraction, or gravitation, appears to be the dominant force. We are told by science that the worlds were formed because vast cloudy masses of cosmic dust began to condense, due to the mutual attraction of each atom to every other atom; and in the condensing process electromagnetic forces were generated. For reasons that science can only conjecture, these great masses began to rotate, and the centrifugal forces caused great masses to be thrown off, which in their turn became small nebulae, with their own whirling motions, until, what with the whirling and the condensing and the throwing off, the suns and planets were formed.

But the work of gravitation had barely begun. These celestial bodies thus formed revolved about each other because of the action of this mysterious force of attraction and momentum. Take the case of our moon, for instance. It is generally conceded that the moon revolves about the earth because it is constantly falling down! We can explain this better by considering what takes place when a boy throws a baseball. If he had merely let it drop from his hand, it would have fallen in a line perpendicular to the earth, but, because he gives it a forward impetus, it falls in a long curve. The greater the forward velocity, the more gradual is the curve before the ball reaches the ground. Suppose, now, that the boy stood upon a mountain higher than any other in the world, and threw the ball with so great a force that the curve downward equalled the natural curve of the earth's surface. Then the ball would parallel that surface forever, provided its momentum remained undiminished. In other words, it would follow an orbital path around the Earth. That, it is contended, is just what the moon is now doing, and what the planets are doing as they follow their orbital paths around the sun. But what causes the forward velocity of celestial bodies? Who knows!

Again, gravity comes into play in that well-known phenomenon called "the precession of the equinoxes." The pull of the moon upon the equatorial protuberance of the Earth causes the Earth's axis to change its direction, so that in the course of nearly 26,000 years it has described a complete circle. Those who have studied this matter tell us that there is a direct correspondence between the astronomical and zodiacal cycles, and the cycles of human evolution on the Earth. As a topic like this is extremely technical, we make only a brief mention of it here.

This mysterious force of attraction is called gravitation when referring to its operations within the confines of our solar system, and a pressure of space when referring to its operations in the outer spaces. Is this universal force — so-called — a property of space? For answer, let us indulge in a little careful analysis of the origin of the universe, and expect the unexpected in our deductions. We shall open our discussion of the beginnings of things with some prefatory comments on matter and energy.

New ideas are constantly coming to the fore concerning the source of the sun's energy. It is believed by science that one of the most subtle forms of radiation, the cosmic ray, is the result of the resolution into simpler atoms of complex atomic structures in some exploding stars. Since matter is indestructible, that small portion of the atom which is left over after it has been resolved into an atom of a simpler structure changes form, and is then dissipated, and manifests as energy. Thus is accomplished the dissolution of a star.

Up to the present time, however, there have been only vague hints as to the possibility of energy manifesting as matter, and so *building up* a solar structure. At any rate, new fields of investigation are opening up before us as we begin to look upon matter and energy as being but two aspects of — space, let us say, provided that we do not look upon that space as something utterly negative and void, which we can describe only by telling what it is not. An article "Space Structure and Motion" by Dr. Gustaf Strömberg of the

Mount Wilson Observatory brings out some interesting ideas concerning the nature of space. Yet, with all the research that is being carried on along these lines, we are forced to ask: Granted that matter is indestructible, appearing, though it may, in changed aspect as energy, how did it originate? The logical answer must be that whatever is indestructible is also uncreatable — which is only another way of saying that all the matter which composes the uncountable stars always existed, in one form or another. Thus we have a doubly infinite universe, endless not only in extent, but likewise in time, at least so far as the materials of which the universe is built are concerned. We are forced, then, to consider two possible deductions: either there have always been stars: for the law of gravitation must have always been in operation, when not latent, or it could not operate now; or else, if this universe once did not exist, save as endless clouds of cosmic dust, then the matter of which the clouds were formed must have been of such a nature that it had not been acted upon by the force of attraction from the beginningless stretches of Duration until the relatively recent period — however far back we choose to place it — when this cosmic dust commenced to form itself into nebulae and suns. This means that the force of mutual attraction, commonly called gravity, is not a force outside of matter at all, *but is a property of matter itself in its present condition.* To rephrase it, when bodies are attracted to each other, that force of attraction is inherent in those bodies themselves, and is not the result of some outside force impinging upon them. It follows, then, that within matter itself there is an energy that is not active at all times, but remains latent for great periods of duration; and that with the once-latent energy awakening to activity, there must be a cause for that awakening. There must have been the need for activity, which is what Theosophy calls the urge to grow. And following this thought farther, there must be the need for cessation of activity after great periods of star formation, else this force could not have once been latent. A force is a force, whether latent or active, and action and rest are its two aspects.

Well and good. We have our choice of deductions, and if we wish to follow them up, we find that they are both true. There *have* always been galaxies of stars, but not the same stars that compose our island universes. So from the standpoint of our present universe, it, as the aggregation of stars that we now have, had its beginnings. But before this universe began, there must have been other universes, former selves of this one. Thus, from a logical standpoint, we must acknowledge that the law of cycles is of universal importance, and that the same causes that brought about the galaxies of today, operated to bring forth other galaxies unthinkably long aeons ago, and that the same causes which brought about the resolution of those one-time universes, will, in the far distant future, resolve our present galaxies into vast clouds of cosmic dust. But as space is endless, there is ample room for the operation of both the constructive and destructive forces, so that it is impossible that there will ever be boundless space devoid of stars. Our whole galaxy will one day vanish with its companion galaxies, but there will always be stars elsewhere. To put it briefly, in this vast expanse of space, there are universes in all stages of growth: (a) universes unborn, the matter of which they will some day be formed being now so tenuous that it is utterly invisible and intangible; (b) universes in the gestation period, in which the cosmic matter has been quickened into activity, and is gradually condensing and concreting; (c) universes in the full glory of their lifetime with the celestial spheres circling in their appointed orbits; and (d) universes dying out, resolving themselves into *invisible matter*, or latent energy. What we are prone to call empty space is filled with universes-to-be. But how can energy exist without matter? It cannot, for it *is* matter itself — a bipolar function of life.

And so we have found that this first chosen course has been a loop in reality and has become one with the other path, in which we recognize the principle of all-pervading life. For what is energy — even when it shows itself as our friend gravity — but a form of life? Let us deviate no longer from our purpose. Let us study the universe as the ancient philosophers studied it, recognizing it as a living organism, composed of living units, the galaxies. To answer the question, then, Whence came the stars? we must say: They grew.

III

The Cosmic Pattern

Great are the mysteries concerning the universe, as taught by the ancient Wisdom-Religion, Theosophy, and greatest of all is the mystery of consciousness or life. The world of science deals with facts — but what are facts? Facts are merely true statements about things, and may, or may not, describe the things themselves. It is traditional among scientists of today to accept only facts — it is the place of religion to reveal the deeper things about which some facts may be discerned. You, for instance, are more real than any facts about you. You, as a factual entity, are but an event in a cosmic drama. And science alone cannot tell you what you *are*, it can only note and describe certain facts about you, endeavoring thereby to gain some knowledge about your true nature. Only religion or philosophy can take up the thread where science leaves off, and tell you what you are. Not until religion walks hand in hand with science, shall we have a true philosophy of life. For our present purpose, philosophy may be thought of as the catalytic agent bringing together these two great modes of thought. Such a true union of thought is Theosophy, for it helps us to see things as they are, and to discriminate between the facts that have a real bearing on those things, and other facts that are indeed facts, but bear no definite clues as to the true nature of the things under study. In this, Theosophy is the synthesis of religion, philosophy, and science, and as such, it not only postulates certain truths, but also gives the reasons for them. Let us now continue the theme of the previous chapter, and follow matter through its transformations into energy — and beyond.

It is declared by some scientists to be a fact that matter may undergo a change, and appear as some form of energy. We have spoken of this in relation to the breaking down of the atoms in certain exploding stars, resulting in the cosmic rays which are so interesting to modern science. And we shall repeat the question: How can an energy be floating about in space, entirely divorced from matter? It cannot, we reaffirm. All matter, we learn, is in the last analysis composed entirely of energy. Study the atom. What were once considered to be hard balls of matter, are now described as whirling energy centers, which, due to their intense vibrations and other motions, give the semblance of solidity. This matter of which the rocks, our houses, even ourselves, are composed is now recognized as manifestations of something called energy, but which Theosophy recognizes as being one form of life. This energy-matter is really a bipolarity. But there are as many kinds of energy as there are of matter, and to adopt the custom of theosophical students, let us call this bipolarity "spirit-matter." This term is convenient in that it will include all the multifarious ramifications of either.

Now, just as north and south are opposite poles of direction, or heat and cold are opposite poles of temperature, or long and short opposites of linear extension, so spirit and matter are the opposite poles or aspects of space. By this we do not mean space as three-dimensional extension, but Space in its archaic meaning, as the carrier of infinite consciousness. In this light, when we observe the transformation of matter into energy, we actually witness the transference of a unit of consciousness from our plane of physical perceptions to a more subtle state, so that its matter pole is sublimated to the point that we call energy. It would follow then, that the spirit pole of that bipolarity will be sublimated so far beyond any recognizable form that it becomes a sort of superenergy, if you like.

This establishes, then, three levels: matter, energy, and superenergy. Let us complete the picture by considering the possibility that consciousness instead of "rising," might "drop" a point, as it were, so that its more subtle energy aspect would be on the level of what we call matter, its material aspect would then fall below our level of perception, and we might easily mistake it for energy again. Actually, we should call it *sub*matter. With our limited percipience we are prone to view anything that is not strictly matter as though it were merely energy, the question as to whether that energy is finer or grosser than our matter being left unconsidered. It is much like referring to those portions of the mind that we do not understand as the "subconscious" mind, about which we know nothing from a purely scientific and biological standpoint, but of which Theosophy declares that while much of it is undoubtedly vegetative, and therefore subconscious, the nobler part is *super*conscious.

To return to our mobile center of consciousness: Wherever it might be situated in respect to our powers of perception, it always carries with it — or shall we say it is always carried by — that bipolarity, spirit-matter. Extending our view still farther, what possible reason is there for doubting that endless scope exists for consciousness above and below our extremely limited plane of perception? This is one of the fundamental teachings of Theosophy.

And now we have a truly unlimited universe. Not only is it boundless in respect to space and time, but also in respect to ranges of consciousness, or levels of energy and matter. In fact, what is energy on one plane might be superenergy from the standpoint of a plane lower, or matter from the standpoint of a plane higher, or submatter from the standpoint of a plane higher yet. And so there is no end. This establishes what, in the Ancient Wisdom, we term "planes of consciousness." Just because we see on this physical plane, stars and planets and other forms of cosmic life, there is every reason to believe, as is taught in Theosophy, that the matter-energy of other cosmic levels or planes gathers itself together, forming worlds on these other planes. It does this through the agency of forces corresponding to those in our own physical world, e.g., the urge for growth. And inasmuch as there is intelligent life on this Earth (from which we may properly deduce the existence of life on other planets), what possible excuse have we for thinking that this physical plane is the only one that can support intelligent life? Theosophy teaches that these endless planes of consciousness provide the appropriate environment for uncounted hosts of *lives*, whose natural dwelling places are the worlds functioning on these various planes of being.

In speaking of these inner planes of being, we refer to them as the subjective planes of life, in contradistinction to this objective physical plane. The word "subjective" has an unfortunate connotation, giving it a tinge of unreality. If a man is said to have had a subjective experience, we are not particularly concerned, thinking that as it was nothing physical he will soon get over it, little realizing that an inner experience can affect and change a whole life. The fact of the matter is that the subjective universe is the *cause* of the objective. For the universe as studied by science is essentially a world of effects, while the universe of causes (more properly the domain of religion, perhaps) is the inner unseen universe, from which flows continuously the stream of cosmic vitality which animates the outer tangible universe. The reality of the subjective universe may best be brought forth in the use of an analogy. Take a book for instance. What is a book? Is it the bound volume, the written page, the style of type used, or the binding? None of these is the book itself. They but form the embodiment of the ideas which are the real book. But these ideas which caused the author to write, the printer to print, and the reader to read, are the subjective book; yet without them, there would be no justification at all for the work which the production of a book entails. Without the subject matter, or the subjective book, there would be no excuse for the objective book having been printed, for it would have been a meaningless jumble of words; a fine-looking volume perhaps, but carrying no message. In like manner, life would have no meaning were it not for the subjective life within us, which is derived from the subjective life of the universe, just as our objective physical bodies are derived from that portion of the objective universe we call our Earth.

Of these things we are prone to say: "But there is no evidence to support such statements. Where are your proofs?" To this, the Theosophist replies: "There is evidence. Use your own methods of studying *facts*, and deducing the truths therefrom. Is it not a fact that you are cognizant of various forms of energy and matter? You cannot see a force, yet you do not deny its existence, for the reason that you are able to see the effects of force working upon matter, and are able to use such forces as heat, electricity, expansion, pressure, and what not. Those forces carry their own proof in what they do. Likewise, those subtler forces of which Theosophy teaches carry their own proof with them in what they do. Will you not examine with us the things these finer forces do, and reserve your judgment until all is said?"

IV

Life Without End

A study of the two preceding chapters has shown us that space may be considered as one with consciousness, and that it is boundless in a number of respects: in infinite extension, in time, in its capacity for containing galactic systems without number — on this, our own physical plane, let alone on any other — and in its numberless modifications of energy-matter, or spirit-matter, which we call planes of consciousness. We have seen likewise that consciousness may shift from one of the planes of energy-matter to a plane higher or lower, causing thereby what *appears* to us as the transformation of matter into energy, and *vice versa*. The logical deduction is of course that matter is indestructible, yet there is a constant exchange and interchange of matter from this, our physical plane, with matter in planes "above" or "below" ours. This is the real secret of the origin of matter. It is part of the reason why we say that a star comes into being by a process of evolution, in the original meaning of the word, to wit, a turning outwards. Much of the matter of which the solar system is built is the residue from previous solar systems on this plane. However, some of the materials were once a part of the systems of worlds on inner planes, and, in making the apparent transformations of energy into matter, and matter into energy, these materials "descended," or maybe "ascended," into our plane of perception, where they were built into a solar system, in obedience to natural law. We nearly said, "in obedience to laws of attraction" — which would not be altogether untrue, but would be very incomplete without taking into account other forces, both material and spiritual.

We have now to consider what causes these transferences of matter and energy from one plane to another. In Chapter III we enunciated the fact that the transference of energy-matter from one plane to another is indeed a shift of consciousness from one level to another, and that the bipolarity spirit-matter is the carrier of consciousness. This will not sound so strange if we reflect on what can be done with electricity. Before there were dynamos, batteries, or even static machines, where was electricity? It was — and still is —a very active force working on a subtle plane surrounding the earth, and interpenetrating it. Without this wonder-worker electricity, the earth could not long endure. But we have the means to bring it under observation and control by merely rubbing a glass rod with a silk handkerchief. The operation of the static machine is very suggestive, for one rotates small lead balls attached to a glass disc, in the vicinity of strips of tinfoil, and the electric charge is picked up by a brass comb under which these lead balls turn, but which they do not actually touch. Here you have, figuratively speaking, motion being transformed into electricity. In this we have accomplished by hand much the same thing that is done by nature during a thunderstorm. Electrical potential is built up between two clouds, or between a cloud and the earth, until it finally breaks through where the resistance is least, and we see a lightning flash; and, to complete the analogy, remember that the electric energy is bipolar, appearing as positive and negative charges of electricity.

Let us apply the analogy, and say that there is a life potential built up between two planes of consciousness, until the stress becomes so great that it breaks through, causing the transference of energy-matter from one plane to another. All phenomena are in reality, manifestations of cosmic life.

Heretofore we have considered only the passage of consciousness from plane to plane in the form of atoms, the building blocks of all things; and have not as yet considered how these atoms may be gathered together to form a living organism; because an organism as such could not be transferred bodily from one plane, or state, or condition to another. When a man dies, his body is not suddenly whisked away to some invisible plane, though the death journey does take the inner man through inner planes of consciousness; but in such a condition of disembodiment, he is, as it were, but an atom of consciousness, and as such he can pass from plane to plane, as will be shown later.

Nor have we as yet touched on the causes which underlie the transference of the atoms from plane to plane. We shall not hold for a moment the idea that cosmic life is woven upon the slim fabric of chance. Behind all phenomena is Intelligence, Cosmic Will, if you like, and as we are about to study the nature of will, we shall now postulate the existence of will everywhere. Cosmic Will is twofold: the volitional will, about which there is a deep mystery, and the automatic will which governs the motions of the heavenly bodies, climate and weather, the flowing of great rivers of cosmic energy, which result in the passage of atoms from one plane to another, and in fact all the phenomena of so-called inanimate nature. But within every organism is reflected the Cosmic Will, which reappears as individual will, both the volitional and the automatic. The volitional will is that which enables us to speak, think, eat, walk, run, and so on, and the automatic or vegetative will governs the natural processes of the body, such as the heartbeat, and the digestion.

All beings, endowed with will as they are, have the power of choice and freedom of action, at least to a degree. Yet they cannot break the ties that bind them to the Universal Self, of which they are a part, nor have they unlimited freedom of will, for cooperation is the strongest force in the universe. When all things act in obedience to the Cosmic Will, they act in obedience to their own highest will, which is the basis and foundation of ethics. The *essence* of the Cosmic Will is compassion itself, some appearances notwithstanding. The law of the survival of the fittest may appear to be the ruling factor on this Earth, but there are higher factors, and we are not to judge the whole of the universe by so small a part as this Earth. Furthermore, there are reasons for this struggle for existence which had their originating causes aeons ago, in other worlds. This however, is not a topic to be dealt with at length as yet. Suffice it to say that there is no nonexistence for that which exists, so that this term "survival of the fittest" is not the best that could be chosen. We shall later enlarge upon the teaching that when a creature dies, it is not thereafter nonexistent, but is taken care of according to its kind until a further opportunity arises for it to return to earth under more favorable conditions. Essentially, the "law of the survival of the fittest" is no more heartless than the sorting out of defective goods in a factory. In the factory the discarded goods are disposed of, but with nature, there is always another chance.

Now, this Cosmic Will is reflected in every atom, controlling to a large degree its motions and progressions from one plane to another. Here then is the picture as far as we have put it together. The atoms have life of a kind within them. Plants, animals, and humans are admittedly alive, and *conscious*. But how shall we fill the gaps between the purely atomic and those organic structures we recognize to be alive; and is man the last word in living things?* Let us consider first, what it is that makes man a living creature.

All attempts to ascribe life's processes to chemical or molecular activities in the brain cells are fruitless, and may be discarded at once. Volition is our first clue, for by an entity's ability to use volition we may ascertain whether or not it is conscious. Volition, or will, is a power shared by all living things, however rudimentary they may be. That which we call human consciousness is but a cross section of an infinitely grander consciousness, which goes by the shabby term "subconscious." Recognizing that the Theosophical ideas about the subconscious differ radically from those current in modern psychology, the time seems ripe to set forth the teachings of the Ancient Wisdom concerning man's true nature, including those portions which embrace the subjective as well as the objective.

You will sometimes hear a Theosophist say that the term "subconscious" is a convenient name for something about which almost nothing is known. This statement will be borne out in due course of time, though we agree that two other terms used today, the

---

*The term "man" is used in this book to denote human kind and, of course, includes women as well as men. Masculine pronouns such as "he" and "him" are sometimes used for ease of expression and also should be understood to include women.

"subliminal" and the "superliminal," give one to understand at least that there are states of consciousness above as well as below that of the normal waking state, centered in what we call the mind.

To begin at the beginning, then, man must be understood as being a cosmic pilgrim who originated in (and has never left) the Boundless. In the beginning of his evolution, he was a life-atom, of which there are uncounted millions filling the spaces at all times. These life-atoms are thoughts, divine thoughts if you will; and are the building blocks of which the framework of the universe is built. Such a building block is man, with an unspeakably grand future before him. If you would form a mental picture of a life-atom, you are not to think of it so much as something composed of matter, like the atom of science, but as a focus of exhaustless energy, altogether without size, shape, or fixed position. Within it are potentialities galore, and its whole future growth depends upon its ability to change its potential energies into kinetic spiritual energies. This is the work of evolution, and the atom slowly learns to use its natural powers, and in manifesting them, to "build more stately mansions," until you have at last that wonder of wonders, man. But the story is not yet told. It is impossible here to trace man's evolutionary course through the various stages of cosmic life, but it should be borne in mind that at all times he is potentially a god, and the human stage in which he now finds himself is but one of the means he employs whereby he may realize and make use of his godlike qualities. How could he be a potential god unless other gods were a reality? He could not, and in this fact you have the keynote of this study. In order to understand man, the gods themselves must be studied. They are divine beings who are such because they have succeeded with relative fulness in bringing into play and active use that exhaustless fountain of spiritual energy of which the life-atom was once the focus.

Now, in reality there are two kinds of gods, inner gods, and outer gods. The fountain of energy locked up within the heart of a life-atom, a plant, an animal, or a human is an inner god. As these same energies pass from the inner or unmanifested realms into the manifested, they become the objective manifestations of a god, and you have an outer god. Obviously the puny body of a man is insufficient to enshrine the holy flame of a god; a body more appropriate must be found, and the gods are equal to the task. They build shining bodies of solar energy; they are the suns. But even this is not the end. There are still unawakened powers within themselves, *inner supergods*, if you like the phrase, and when these inner supergods become outer supergods, their bodies are the galaxies themselves. Actually, there can be no hard and fast delineation between that which is inner and that which is outer. Both are relative, as we shall see shortly.

There is no such thing as an entity existing on one plane alone. Because all things are composite, the various elements and substances forming things are derivative of the various planes of consciousness. For instance, my physical atoms are derived from this physical earth, the astral atoms forming my astral body are derived from the astral plane, else they would not be astral atoms.* The spiritual atoms, forming in their aggregate the vehicle of my spiritual forces, or buddhic principle, originated in the spiritual plane, and so on through the entire constitution.

---

*"Astral" is the word used to denote the plane of consciousness composed of energy-matter one grade finer than that of this physical plane. The astral body of man is the pattern around which, atom for atom, the physical body is built, and grows to maturity.

Moreover, from my viewpoint, my spiritual and divine "principles" are the inner part of my constitution, and my vital-astral-physical nature is the outer part of myself. That portion of my constitution wherein my consciousness is centered, in other words, my mind, is neither inner nor outer. If I were able to center my consciousness in the spiritual portion of my being, then my mind, as well as all that I now recognize as outer, would be exterior to me, and my spiritual nature would be neither inner nor outer, though my Divine Self would still be inner. So this question of the inner and outer is a matter of relativity, as are all so-called opposites. It seems quite reasonable to consider the subconscious as being the inner, as contrasted with the outer. And because the powers and faculties manifested in the outer part of the constitution are well-nigh insignificant compared with those latent or partially active in the inner reaches of ourselves, we find, much to our astonishment, that people who through unusual circumstances go into trances sometimes give evidence of having knowledge that far transcends anything they have learned in books, for they have temporarily tapped the reservoirs of the subconscious. When we realize that this is in reality the inner self, we see at once how utterly inappropriate is this term "subconscious," for it tells us nothing.

Let us now apply this philosophy on a grand scale. It was observed that the atoms of which our constitutions are composed are derived from the various planes of consciousness. Let us not get the erroneous idea, however, that the universe exists for nothing more than the building of the principles of man. Important as this process undoubtedly is, it is merely incidental to the real activities that go on in these inner planes. Actually these planes of consciousness are the principles forming the constitution, inner and outer, of a Cosmic Being so ineffably grand that it is utterly beyond our feeble powers to define it. And all the hosts of beings who draw their life from the life of the cosmos are merely atoms, building blocks if you like, of which these cosmic principles are composed. Think of it! As you sit at your desk, fully conscious, using your physical and intellectual powers, thinking your thoughts, uttering your words, you are actually in the core of your being a part of the inner constitution — or possibly outer — of some divine being. Similarly, a portion of your inner nature may be the outer universe from the standpoint of the infinitesimal lives composing your inner self.

Why is it that we are unable to define the nature of this Cosmic Being? An incomplete answer would be: because it is so grandiose, so supernal in essence, that words would utterly fail, even if we did intuitively catch some adumbration as to its true nature. A more complete answer would be that there are departments of life that do not lie within the sphere of our understanding. All beings have two aspects of their lives, the voluntary and the involuntary or automatic. The most obvious portion of our own lives is the voluntary, and it is only by study and observation that we may follow the intricate working of our automatic life processes, such as breathing, heartbeat, and digestion. Think of the marvel of the construction of the physical body. What divine powers guided the building of the human eye, for instance, or the wonderfully complex brain? Now, there is a marked contrast wonderfully complex between our view of man as a volitional being and our understanding of the universe, and particularly our solar system, which we hold to be the outer garment of a Superdivine Being, a Solar Consciousness. The aspect that we contact is the automatic, and we find it utterly impossible to receive even the remotest impressions of the volitional will of the Cosmic Being of whose inner nature we are a part. In other words, what we see as the phenomena of nature, the motions of the heavenly bodies, weather and climatic changes, and the like, are simply manifestations of the automatic will of the Cosmic Being whose embodiment is the universe. It is not to be thought that we are speaking of a personal God, for if we think of it as personal, we merely invent attributes and attach them to a being which confessedly we are unable to understand. Moreover, we are safe from the limited view when we reflect that Theosophy postulates no god so high that there is none higher. This Cosmic Being of which we are speaking is but one of many, each manifesting as a universe, and these are legion. Now the natural laws which prevail in any given universe are merely the workings of the automatic aspect of that universal consciousness.

A master of life is one who is able, not only to draw upon his own resources, inner and outer, but through the medium of his own inner constitution, to contact the inner planes, and draw upon the resources of the Divine Being whose "principles" are these "planes of consciousness," and therefore a part of whose nature he himself is. And these resources are exhaustless. The idea is not untenable that sooner or later, with constant effort and growth, the powers of the initiate can become so great that at last he attracts the attention of the Divine Being, and then new and supreme initiations are in store for him. He then passes out of the realms of illusion, into the unmanifested or subjective realms, which, when entered upon, prove to be Reality itself, for he has contacted the volitional aspect of the life of the Divine Being.

We will all agree that man as he exists today does not represent the acme of possible development. He is slowly working out his destiny and will ultimately reach the goal of human perfection. Such perfection, then, must be at least a potentiality, which it could not be were it not also a latent reality. And this teaching of human perfectibility leads us on to new ideas which we shall explore presently, though we speak of it here in order to bear out the premise that there are states of evolution higher than that represented by average man. And there are persons of exceptional development who are in a position to cognize higher intelligences, which again are so lofty in their degree of development that they are utterly outside this plane of human life; and we must call them gods, as a generic term, for of these there are endless grades or classes. From this point on, can we not speak of "God" as a collective noun, meaning gods, just as "man" is used to denote the whole of humanity?

V

What Is an Entity?

We are learning now that the universe is filled with living beings, and they may be classified, in accordance with the incomplete knowledge that we have gained thus far, into three kinds: (a) the simple, or atomic; (b) the complex, or entities comprising the human, animal, and vegetable kingdoms, having bodies composed of myriads of atoms; and (c) the apparently bodiless elemental forces of which we have previously spoken, as well as the higher intelligences which we called the gods. These, so far as we have just studied, have no material bodies, though we must acknowledge that, as energy and matter go hand in hand, there must be some kind of corporeal vehicles in which these beings function, which counterbalance their energic aspects. Their corporeal vehicles may be quite invisible to us, not so much because they cannot be seen, as because we do not recognize them for what they are. Again, in certain cases they are in fact quite invisible, and these we may study by inference and logical deduction rather than by actual observation.

For lack of the right perspective, we are apt to think that a thing is alive by reason of its having a functioning physical body. In time we shall learn that the physical body is the least permanent of all the parts or principles that go to make up any entity, and is employed by the entity for the purposes of experience during its stay in a world such as ours, in which knowledge is gained through the use of physical senses and faculties. To gain a more complete understanding of what an entity really is, we must look upon the energic side of its makeup as being by far the more important, for it is the seat of the consciousness that the being is in its essence.

Consider with me then that an entity, regardless of its class, is embodied consciousness. It is doubtful whether we shall ever find words to describe consciousness pure and simple. Quite superficially we might call it the awareness of sensation, internal and external. Again, we might call it self-awareness, but that is not descriptive of the Self. Perhaps to say that consciousness — or "THAT," as we call it, for lack of a better word to describe it — is the vital drive of the universe, is as complete a word-picture as we can give. The consciousness within a human, or animal or in any kind of being whatsoever, is an inherent power originating in THAT.

By way of analogy, let us suppose that we are looking at the reflection of the sun in a clear pond. We know that the sun is the source of the light that is reflected from the surface of the water, and that whatever happens to the pond will in no wise affect the sun. A breeze ruffles the surface of the pond, and we find that the reflected sun has been broken up into numberless fragments by the ripples, each one a complete but distorted miniature reflection of the sun. The one has become the many. Let the pond represent space, the great reflector of THAT which is represented in our analogy by the sun. When space awakens into activity, this reflected consciousness breaks up into numberless multitudes of fragments which are the consciousness sparks we call entities. At the close of the cycles of activity, space sleeps — the pond becomes quiet again — and the images return into the one reflection of THAT. Through it all, THAT has remained calm and unmoved, though being the source of all its energies.

Any entity is, then, a spark of THAT which has entered Space, or Cosmic substance, modifying it, and thereby forming for itself its bipolar vehicle, spirit-matter. This is a general statement of truth, and applies to any entity, its grade or degree of development determining merely to what use it can put its powers, and build for itself a gross or subtle bipolar vehicle, which would place it high or low in the scale of planes of consciousness.

In support of these remarks, let us consider what a human being is. To ask first a rhetorical question: can a man own something and be himself, as well as being the thing he owns? Assuredly not, you will say. Very well, then. A man speaks of "my hands," "my feet," "my head," implying possession of these things. Is he then his hands, his feet, and his head? By no means; these are instruments that he has created for his use. He speaks further of "my body." Is he then the body? No, the body is a very complex instrument, part of himself indeed, built for his use while he is here on earth. The man is something finer than the body. He speaks of "my mind," meaning possession again. Thus, he is not his mind; he uses his mind as a finer instrument than the body, but an instrument nevertheless. Then he speaks of "my soul." Thus, the man is something more impalpable than the spirit itself. Again, he speaks of "my divinity," implying thereby that even divinity is a means whereby he gains cosmic experiences, so that he is something more sublime than divinity itself. What is the very core of his being? THAT alone; nameless; not to be known by any multiplication of words, but to be cognized as himself. Knowing THAT, he shall one day know all, for THAT is all that it is possible for him to know as man.

And what is it that determines whether an entity is of high or low degree? It is its greater or lesser awareness of THAT as its essential self. The "gods" have become such because they have succeeded after aeons of evolution in making their energic material vehicles more completely representative of THAT. The lower entities which function mainly under the direction of what to them is their higher involuntary wills, as the animals do for instance, represent as yet very imperfectly (because automatically and not self-consciously) THAT. And here is a paradox. One usually feels closer to the heart of things when alone with nature than when in the company of fellow beings, evolved and intelligent though these may be. That is because divinity is mirrored undistorted by the hills, the trees, and the wide expanse of sky and sea. There, however, divinity is automatically reflected, while man is in the peculiarly difficult position of being an entity that is learning the use of his own self-conscious divine will. When he learns to do consciously and with full intent what nature intended that he should do, then he will see each human face as a "mirror of infinite beauty."

There is now another step to be taken in the understanding of what an entity really is. In speaking of man as "possessing" body, mind, soul, spirit and divinity, we are in reality enumerating various gradations of the bipolarity which is the carrier of his consciousness. While we have as yet by no means given a complete picture of the various phases of consciousness which go to make up the complete human, we may use the terms already employed with a certain few modifications and additions, and make up a paradigm.

	Energy	Matter
Divinity	7	1
Spiritual Soul	6	2
Mind	5	3
Emotions, Passions, and Desires	4	4
Vitality	3	5
Pattern, around which is built the	2	6
Body	1	7

This shows at once that man is sevenfold, or, in other words, a bipolarity composed of seven lesser bipolarities, each one of which is in itself an *entity*, a bipolar carrier of its own consciousness. The terms used here are the English equivalents of the original Sanskrit words as set forth.

Atman	Divinity
Buddhi	Spiritual Soul
Manas	Mind
Kama	Emotions, Passions, and Desires
Prana	Vitality
Linga-sarira	Astral pattern, around which is built the
Sthula-sarira	Body

This table will not be so formidable, perhaps, if we show how the septenate may be classified to fit the more familiar division of man as a trinity — body, soul and spirit, after the manner of St. Paul.

Atman	Divinity	Spirit
Buddhi	Spiritual Soul	Soul
Manas	Mind	
Kama	Emotions, etc.	Body
Prana	Vitality	
Linga-sarira	Astral pattern	
Sthula-sarira	Physical Body	

This division of man's nature into a septenate is not new. It may be found in various forms in some of the oldest religions of the world. But let the student keep in mind the underlying principle behind whatever system may be studied, namely that each grade in the constitution of man is in itself a bipolarity, which means that each one is in itself a carrier of consciousness; so we see that man is not one entity, but is composed of a number of complete entities! Each of these is a living, growing thing, with its own life, yet subservient to the main stream of consciousness that we call the man. (The student is here advised to become acquainted with the use of the term "subservient entities," as it will be used often. It refers to the many lives which, though having their own evolutionary pathways to tread, yet do so under the spiritual guidance, so to speak, of the central stream of consciousness, which in this case is humanity, though it may be anything ranging from a sun to an atom.)

Now that which is going to determine the type of person an individual may be is the choice one makes of a subservient entity in which, or through which, to focus his own stream of consciousness. The man of today should focus his energies through the higher mind, or *manas,* but unfortunately he is too often centered in his emotions and body. The master of life is one who has succeeded in raising his consciousness so that it focuses in the spiritual soul, or buddhi. The god is one who has focused his consciousness in his divinity; and then he passes out of the human sphere of cosmic life, and enters a new plane of being, which explains why the "gods" are to us invisible. As such, a god by no means loses his sevenfold constitution, but all the entities which make up that constitution are raised to the higher level, and — well, that is another story.

But if an entity can be a fully sevenfold being on another plane of cosmic life, it follows quite properly that all entities, regardless of kind, are sevenfold entities, so that we can no longer distinguish between "simple" and "complex" forms of life. And to follow the thought still farther, since every entity is sevenfold, it is to be deduced that each and every one of the seven entities making up a human or any other constitution is in itself a sevenfold being, composed in its turn of living entities, each of which again is sevenfold, and so on without end. Not only space is endless, but we learn that even now, every entity is an infinity of lives — a hierarchy.

VI

Microcosm and Macrocosm, the Doctrine of Hierarchies

A true analogy must be one that has universal application. The paradigm used in the preceding chapter, showing the various gradations of energy and matter which compose the inner and outer constitution of a human, must apply in principle, though obviously not in detail, to *any* entity. In order to follow this theme, we shall discard names, and make a new paradigm consisting now of seven lines, and the various values of energy and matter.

ENERGY	MATTER
7	1
6	2
5	3
4	4
3	5
2	6
1	7

The vertical line running through the paradigm represents the stream of consciousness which clothes itself with this complex constitution. Such a figure may now represent *any entity whatsoever,* be it an atom, a human, or the highest god of which we can conceive. It may therefore represent any one of the "subservient beings" mentioned in the preceding chapter — beings which are the units represented by any one of the bipolarities illustrated by the paradigm. That each one of these must have its own subservient beings postulates at once one of the fundamental teachings of Theosophy, namely, that the universe is built upon a hierarchical plan. It may be stated in this wise: *Any entity is composed throughout of hosts of lesser entities,* and conversely, *any entity is but one of a host of like entities composing a greater entity.* Applying this universally, we are forced to admit that there can be no such thing as an ultimate "highest" or "lowest" in the boundless universe. We may think of a Supreme Being watching over humanity, and there is such a Being, as we shall later see; but he is only one of hosts of Supreme Beings, watching over millions of groups of evolving beings, some higher, some lower, and some on a par with our own humanity. And all of these numberless Supreme Beings are subservient to, or watched over by a Super-supreme Being, who is only one in *his* class; and so on endlessly.

Now let us consider what such a Supreme Being might be like, and let us take for an example the one that watches over the entire solar system. Such a Supreme Being is a sevenfold entity, as are all entities. But its seven elements (all of them bipolar, as shown in the paradigm) are what appear to us as the seven planes of consciousness, already mentioned when we dealt with the matter of the atoms being transferred from one realm to another.

Now we live on one of those planes of consciousness; in other words, we are subservient entities within one of those beings which in turn is subservient to, because a part of, that Supreme Being. Say that that Being of which we are a part is the Intelligence working through the sun; then we are the atoms in the constitution of that solar entity. Very unorthodox, some may cry, yet it is not only logical, but it does explain the operations of the universe.

We shall pass once more to the constitution of man. In principle, the laws governing human life are the same as those governing the life of the universe. We now understand better the meaning of the symbolic statement that man was created in the image of God. Just as the seven elements of the Supreme Being are the planes of consciousness upon which are functioning so many forms of cosmic life (among which we humans are included), so the seven elements of the constitution of man, known as the seven principles, are in reality seven planes of consciousness upon which and in which are living unnumbered hosts of beings (life-atoms, as we call them). These life-atoms find in their planes of consciousness (our own human principles) just the kind of environments that they require in order to acquire experience suitable to their relatively unprogressed development. And they pass from plane to plane, just as we have seen that the cosmic stuff of which the stars are composed passes from cosmic plane to cosmic plane. In due course we shall follow this thought farther where it relates to man's progress from one plane to another, though we might say in passing that a typical example of the passage of a human consciousness from our plane to another is what we call death. The purpose of this present chapter is to show that man is a microcosm, doing exactly in the small way what the macrocosm, the universe, is doing. And more than that, so far as the entities composing the man's constitution are concerned, he is already a universe of a kind. To put it mathematically, man is but one term in an infinite geometrical series, the other terms of which are the endless types of beings above and below him.

Now, there are many instances where a teaching is better given by the use of analogy than by a studied attempt to set forth the various factors involved in textbook language. The use of analogy owes its value to the fact that the laws of the universe are fundamentals of natural life. A law, then, may have applications in a number of ways, and instances of its operations, though seemingly unrelated, actually show the workings of that common law, and may therefore be studied together.

We may carry this thought a step farther, so that in addition to using analogies for the sake of illustration in order to drive home certain salient points about the doctrine, we may now study the workings of the laws as such. Applying the maxim "as above, so below," we find that there are things in nature whose relations to one another are more than mere analogy because they are actually one and the same thing. Because the principles of man are derived from the planes of consciousness, the laws whereby the former operate must likewise hold true of their counterparts, and vice versa.

The relation of the human principles to the planes of consciousness is a *correspondence.* White light breaks by a prism into the seven colors and reunites into white light once more by means of a second prism, illustrating how the one light of consciousness becomes sparks of individual consciousness and returns to one again: this is an *analogy.* Analogies may be found everywhere and quickly perceived by the mind of the student seeking for the explanation of a teaching; correspondences, on the other hand, are by no means so easy to discover, for they form the very framework of the universe. As one pursues studies and penetrates beyond the exoteric into the esoteric, the significance of correspondences grows, and one soon realizes that the keys to practical occultism are to be found in the Law of Correspondences. For our present purposes only one line of correspondences will be considered, that dealing directly with the constitution of the universe on the one hand, and of the human being on the other.

Toward this end, let us view the constituents that go to make up the being we call a human. First of all, the body is the most obvious part, being the most objective. The first thing we notice from a study of the human body is that all bodies show marked similarities, so that a surgeon finding a certain bone in one person knows that he will find a corresponding bone in all others, and so on. In other words, we see at once that the human body is built after a definite plan. This plan we call the "astral pattern," and it is inseparably linked to the body throughout a person's life. They are kept together and sustained by the vitality of the person, which postulates a third factor. These three, the physical body, the astral pattern, and the vitality, function as a unity, and we call it the "vital-astral-physical" portion of the constitution. Behind all this stands the desire for life. This desire is the driving impulse, and can be good or bad, depending entirely upon the individual. In some it may be the instinct of self-preservation only; in others, the desire to acquire, to get for oneself; in others again, ambition, the desire to become greater than one's fellows. In its highest form it is aspiration, a desire to reach upward in thought and feeling to the higher planes, and to know one's divinity. However that may be, desire is a very potent factor in the human makeup. Let us then set forth these four "principles" in tabulated form:

Desire
Vitality
Astral Pattern
Physical Body

and consider the source of each. Every part of our constitution has a cosmic source. The cosmic source of the physical body is the earth, for it is from earthly materials that the body is built: the food we eat, the air we breathe, etc.

The cosmic source of the astral pattern is the astral light, as it is called, surrounding the earth, interpenetrating it, and extending for a considerable distance above and around it. It is the great reservoir of human thoughts and deeds. Its power is great, though little suspected. Diseases, war, and other calamities issue from the astral. A war does not originate on the battlefield. It starts in men's minds, which are bathing in the astral light all the time, and continually pick up the mental dross that is left there by polluted minds. How much needed is a positive attitude of mind, so that men and women can raise themselves out of the lower astral realms, and into the sunlight of the spirit beyond! Then the mind will be capable of entertaining bright visitors in the shape of noble ideas, constructive impulses, and the compassionate desire to work for the upliftment of the human race.

The vitality of a person has its cosmic source in what is called by Theosophists cosmic electricity. This might be best thought of as cosmic vitality, animating the whole of the universe, and appearing to us in many guises. One of its lower forms is electricity itself; another, radiant energy; another, mutual attraction (called gravitation); another form is motion. All these are aspects of cosmic life. This flows through us all the time, and we call it "vitality."

Desire has its source in the cosmic urge to become. It is that which brought the worlds into being. What a tremendous urge it must have been to cause the galaxies to grow!

Now, the human principles as named thus far might equally well have been applied to an animal; but we come now to a consideration of those finer portions which give man his proper place in the cosmic scheme. That which makes a person human is the thinking faculty, that in which arises the feeling of "I am I." This we call the mind, and its cosmic source is Cosmic Mind, in which the modus operandi of the universe is first conceived. Innumerable hosts of monads compose this Cosmic Mind, a mystery truly. One class of these monads, which we call human, embodies itself in the principles as outlined above.

Now the mind wavers continually between the Spiritual Intelligence which is the sixth factor or principle (as mind is the fifth), and the desire principle. It might be well to point out that intellect is a technique of the mind, whereas intelligence is a quality of the spirit. The mind devoid of spiritual intelligence becomes coldly intellectual and is likely to be run by desires, and as such becomes an entity which we call the lower self. On the other hand, when the mind reaches upward and unites itself with the spiritual intelligence, it becomes the higher self. As such it is capable of infinite growth and spiritual freedom, as contrasted with the bondage of the lower self.

Now, this Spiritual Intelligence, or the Spiritual Soul, is a direct emanation of the Divinity within, the Inner God. This Inner God is a spark of Cosmic Deity. Deity in this sense does not imply an extracosmic God. Such a being is a logical impossibility. For to make God the cause of the universe, and then set him outside of it, is to separate the cause from the effect, and this cannot be done. Cosmic Deity means the whole of the subjective universe, which itself fills boundless infinitude.

Let us then set forth these cosmic principles, and their counterparts, the human principles.

Deity	Divinity, the Inner God	
The Gods	Spiritual Intelligence }	Higher Self
Cosmic Mind	Mind } . . . . }	
Urge to become	Desire } Lower Self	
Cosmic Electricity	Vitality	
The Astral Light	Pattern	
The Physical Plane	The Body	

It often helps to bring home a doctrine more forcibly if we make a diagram, and to this end, let us think of the essential ray of THAT, before alluded to, as the very root of all that a human is, as though it acted like a lens, having the power to bring the cosmic forces to a focus whenever there is the need for embodiment. Just as a burning glass may be used to throw an image of the sun, which is a real thing, composed of sunlight, and which is capable of setting fire to a sheet of paper, so let us think of this image of the universe as a real thing, such as we know a man or woman to be. We can then more readily grasp the meaning of the teaching that each and every one of the human principles is derived from a cosmic plane, or principle, or element. These seven cosmic forces will correspond to the seven colors which make up what our eyes tell us is white light, for we know that the image of the sun is capable of being separated into the seven prismatic colors. This particular illustration can be useful in that it impresses upon our minds the fact that the seven principles, whether considered as parts of the universe, or of man,

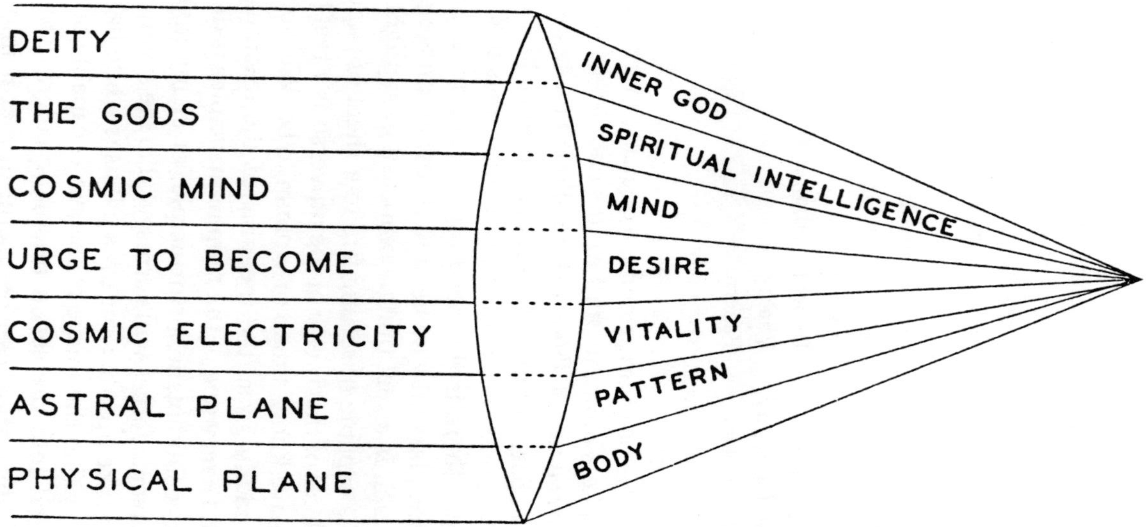

DEITY
THE GODS
COSMIC MIND
URGE TO BECOME
COSMIC ELECTRICITY
ASTRAL PLANE
PHYSICAL PLANE

INNER GOD
SPIRITUAL INTELLIGENCE
MIND
DESIRE
VITALITY
PATTERN
BODY

are not separate things, which may be pigeonholed, but that they all interpenetrate, just as the seven colors are all blended together to make white light.

It may now be made clear that life and death are not opposites. The opposite of death is birth, for each is a gateway in life, one leading into and the other away from the objective world we live in. Death, in our diagram, would be represented by the withdrawing of the lens, so that the cosmic forces remain as cosmic forces, and the image disappears. And here is the importance of the teaching of the duality of man in respect to his having a higher and a lower self. If a person is capable of uniting with the spiritual soul during earth life, he naturally rises with it to its own Cosmic Source, and experiences more or less fully, depending on the evolutionary standing of the individual, the events which make up the cosmic journey which is undertaken by the spiritual soul. Even when unaware of the inner planes through which it is carried by the spiritual and divine Self, the human consciousness is dreaming lofty spiritual dreams in a condition which is called in Theosophical parlance the Devachan. This may be experienced anywhere, for the journey undertaken by the spiritual-divine entity takes one far afield.

If, on the other hand, the motif of the life just lived has been self-aggrandizement, self-indulgence, indifference to others' sufferings, the giving way to lust and passion, then one is united with the desire principle, and the entity functions as the lower self. As such, he is not made of the caliber that can stand the rarefied air of the spiritual planes, and he sinks into the lower planes which are the cosmic sources of the lower portions of his being. These lower planes are centered closely around the earth itself and are the astral light in its lowest form. There the entity must remain for a time more or less long, depending entirely upon the life just lived, and a period is spent in trying to dissociate himself from the chains of his own forging. This condition is known as kama-loka or the place of desires.

Imagine what the entity must suffer, desiring strongly, yet having no means of satisfaction. Here, it might be pointed out lies a grave danger in all matters pertaining to spiritualism. For the entity seeks, and often finds, satisfaction through the medium of the séance room. This reacts most unfavorably both for the medium and for the discarnate entity. The latter's stay in kama-loka is prolonged thereby, and his karma is linked with that of the medium, which has to be reckoned with in a future life.

What a need there is for more light on these subjects! The pity of it is that in just those quarters where the teachings are most needed they are spurned, and so humanity goes on suffering. When will we wake up and see the meaning of life, and the great responsibility that rests upon our shoulders as men and women of a highly complicated age? Evidently we have not yet learned all that suffering has to teach.

VII

The Laws of Nature and the Seven Necessities for Growth

Did you ever go out alone to greet the sunrise, and feel that the radiant warmth and light of the sun and the crimson and gold of the clouds above the pastelled mountains had a special message for you? Have you never caught the conversation between the pine tree and the wind that comes from the east bringing tidings of a storm? Have you never mused on the contented music of the stream absorbed in the business of getting to the sea, and the silent music of the old brown hills that reach in vain for the sky? What, O hills, is this magic, that it stirs us to deep questioning?

And the answer comes — Beauty. And what is beauty but the recognition of the life eternal which is reflected in all things. Had we not this life eternal we could not sense it elsewhere. The only possible interest we might take in a sunrise would be to study the various angles of reflection in the drops of water in the clouds which cause the predominance of the roseate light. For the intellect working alone is cold and unfeeling, but when illumined by the splendor within, it sees and recognizes the splendor without.

But it is only natural that we should want to know

the reasons for the various wonders that we see in nature. Why does the wind blow? Why do the birds migrate? Why does the Earth rotate? Where did the moon come from? And so on and on and on.

Now, there are two kinds of laws of nature, the so-called purely physical, as for instance those dealing with velocities, pressure, inclined planes, weights and measures, planetary motions and measurements, and what not — all those laws which may be expressed in mathematical terms; and those less rigid, we may say, which manifest as the instincts of plants, animals, and humans. All things obey the "rigid laws" of mathematics, though only "living creatures" come under the influence of the instinctual laws. In either case, we must look for the root of the laws of nature in the automatic will of the Great Being of whose constitution our physical plane is but one of the elements or principles. From the fact that the physical laws seem to apply throughout our Milky Way, it would seem that this whole galaxy is the outward manifestation of that Great Being. We might do well to look for the seat of the instinctual laws as known to us in the automatic will of that nearer Supreme Being we spoke of in our last chapter, who functions in and through our sun.

What, then, of the volitional wills of all of these Supreme Beings previously mentioned? As yet we seem to have contacted their automatic wills alone. That is true; and it should be stated here that within the knowledge of man, there is no god, however high, who knows or understands the workings of the volitional will of the Great Being who embodies itself in our greater cosmos composed of island universes. It is probably just as impossible for us as yet to contact the mighty thought processes of the Cosmic Being as it would be for an electronic intelligence in our little finger to inquire into the intellectual and spiritual processes of our human intelligence. In either instance the one is so far outside the plane of the other, that, save for the automatic flow of the life currents, there is no conscious contact between the two. The volitional wills of these mighty beings, however, are not without their effects upon us.

The heart of the universe is compassion itself, and that within ourselves which is divinest, which is the seat of our inspirations and our aspirations, and which recognizes the beauties of nature, which feels love for all things, is in constant and active communication with the heart of the universe, which latter is the seat of the volitional as well as the automatic will of the (for us) Supreme Being in the cosmos.

And now, to take up the second theme of this chapter, we must once more hold before the mind's eye the picture of the universe filled with intelligent beings of all kinds, all working together in hierarchical manner for the common good. Let us think of any entity as a complex bipolar manifestation of consciousness. Such an entity, wherever it may be, is a growing thing, pursuing a cosmic journey, which will ultimately lead it to the heart of the universe itself. But there are certain means whereby it may achieve its end. These means are seven in number and may be called the seven necessities for growth. Technically they are called "the seven jewels of wisdom," because they embody the whole of the technical aspect of the teachings of Theosophy. They are not arbitrary doctrines, but may be thought of as seven properties of consciousness, just as we think of weight, volume, mass, etc., as properties of matter.

At the outset of our studies, we understood the logic in the periodic coming into being and the vanishing of the universes. In this we saw an instance of the operation of the first of these necessities: periodic reembodiment. This is a law which applies to all beings. In order to grow we must act, but in action there is fatigue, and rest is necessary before action may be continued. In order to act, we must find the means to act, and the finding of the means is embodiment. Rest, then, is disembodiment, and the resumption of action is reembodiment. And since every embodiment is but the resumption of action, the first necessity for growth is reembodiment. (We might make it clear that the teaching of reincarnation is but one phase of the general law of reembodiment. Reincarnation means "re-enflesh-ment," the taking on again of a body of flesh, and as such it affects only man and the animals. All things reembody, regardless of the types of body they assume.)

Since the impulse to reembodiment comes from the consciousness center itself, the consciousness center is therefore building for itself just the kind of vehicles that it requires at any time, in order that it may manifest the innate powers and faculties that it has developed up to that time. All beings have untold latent possibilities for future growth, and the vehicles they build will always be suitable for the work they have to perform in manifesting or representing the awakened powers of the entity. In the following sense we learn that every entity is self-made: at any moment it is the sum total of all that it has been since the beginningless beginning, and at the same time it is always the cause of what it will be in the future. Every entity is, then, its own law of cause and effect. There is a technical word which describes this law, which is now no longer strange to Western ears, and that word is "karma." It is derived from the Sanskrit root *kri,* meaning "to do." Therefore it is the law of action and reaction, or cause and effect. Each entity is its own cause and effect, and therefore entirely responsible for itself. Karma, then, is the second necessity for growth, or another jewel of wisdom. The karma of an entity determines the plane of life that it will occupy. An entity of a simple or undeveloped karma will occupy a grade suitable to its needs, and it is automatically drawn there by its karma.

Further, the universe may be likened to a great ladder of life, the rungs of which are the various classes of entities in their various stages of being. This is the third jewel of wisdom and is the doctrine which explains the hierarchical structure of the universe. Inasmuch as the entire universe is composed of lives, and as each life has countless lives subservient to it and helping to build its inner and outer vehicles (take for example ourselves, our own bodies are built of cells), so this doctrine of hierarchies may be summed up as was done in a previous chapter: *Every entity is composed throughout of lesser entities*, and conversely, *every entity is but one of a host of like entities forming a greater entity.* Contrast this with the so-called law of life on this earth where every creature is supposed to live for itself alone!

Now there are uncounted hosts of beings occupying any particular rung of life. Consider members of the human family, for instance, as being all more or less on the same rung, because we are all humans. But there are no two identical entities to be found anywhere. It is illogical to think of any two beings as exactly alike, for if there were, they must have followed identical lines of evolution since the dawn of time, and must therefore be acted upon by identical influences, not only subjective but objective, which means physical. This could only happen if they occupied the same space, which of course is impossible, for then they would be *one thing*. The individuality that makes every entity differ from every other one, and react differently from every other one, is its own essential characteristic. This we call its *swabhava*, a Sanskrit word meaning "self-becoming." There are billions of humans in the world, but the kind of human that each one is, is the expression of his swabhava. Thus, the fourth necessity for growth of any entity is its essential characteristic.

These four necessities may be epitomized thus: *reembodiment* is the process whereby an entity acts, and then rests, assumes vehicle after vehicle for the purpose of making its own *karma*, which places it high or low in the *hierarchical* structure of the universe; and the manner in which it fills its place is the expression of its *swabhava*, or essential characteristic.

The fifth necessity is that the entity bring out its latent faculties and powers, else it can never grow. This is the process of evolution. Later we must consider at length some of the facts concerning evolution, but here let us point out that the word means more to the Theosophist than to the layman. To the Theosophist evolution is not alone the change in the form of a creature to adapt it to changing environments. It is not a process controlled from without. It is mainly growth due to an inner urge to grow. Etymologically the word itself means "turning outwards," and as such it is understood by the Theosophist to mean the bringing to the fore, or into activity, of hitherto latent powers and faculties. That the increased powers must find their means of expression is undeniable; and we learn that the body or form changes as the need grows for expressing or representing the ever-growing entity within. The Model T Ford did not grow into the Ford V-8 by a process of adaptation to environment; the mind that once conceived and built the Model T, later built the V-8.

There is a phase of the doctrine of evolution that will be brought out more fully later, but we mention it here because it is important to us as human beings. Evolution in its usual course is a slow process, but man has it in his power to take it into his own hands and evolve with greater speed than the entities in the "lower" kingdoms, and it is due to this fact that we find history lighted here and there by the lives of exceptionally great people who have been a real power for good in the world. Through a series of lives they have blazed a trail in advance of the rest of mankind, setting an example that all of us can follow. It is just because we have the awakened use of mind that we can become great, that we can speed up our evolution if we choose to do so.

These last words have already enunciated the sixth of the great necessities which is called the path of immortality. This, and the seventh necessity, self-knowledge, are perhaps the most important, and certainly the most beautiful of all the jewels. While evolution is the process of bringing out the latent potentialities of an entity, the various stages of growth passed through as the result of the awakening of these powers is like a great pathway in which the individual is at once the traveler and the path. One does not have to go to the far corners of the universe to become a god, he becomes a god, wherever he is.

If he develops his godlike qualities, man then reaches, because he *becomes,* his goal. And what is his goal? Self-knowledge, the seventh necessity. This is the *comsummatum est* of all our endeavors: that we should know and *be* what we are in the core of our being. Because each one is a spark of the cosmic consciousness, a ray from THAT, it follows that when we have evolved to a union with THAT, and have grown to *be it*, then there will be nothing in the spaces of Space that will be hidden from our gaze. We shall have relative omniscience, because we shall realize that we *are* one with THAT which is omniscience itself. These, then, are the seven necessities for growth.

VIII

The Making of a World

A common nail wrapped around with insulated wire provides an interesting illustration of the manner in which the universe is quickened by the life-energies pulsating through it, for when an electric current is passed through the wire, the nail becomes magnetized. The north and south poles of the magnet represent the two poles of consciousness pervading the universe and manifesting as spirit-matter. Remember that both spirit and matter are forms of energy. Now, withdraw the nail from the coil of wire, cut it in half, and reinsert the halves, keeping them separated, allowing the current to flow through the wire. By testing with a compass, you will find that each half of the nail is a complete magnet in itself. Cut the nail into quarters, and repeat the experiment, and you will find that each section of the nail is a magnet with its two poles. Imagine that you have cut the nail into so many pieces, that each piece is but one molecule of iron, and the same fact will hold. Each molecule, and indeed, each atom of the iron in the nail is a complete magnet. In fact it is held that every atom of the iron in the nail is a magnet all the time, but that the reason the nail as a whole is not a magnet until you pass an electric current around it is that these atoms are placed higgledy-piggledy, pointing in every possible direction, so that there is no unity of magnetic action. The electric current rearranges these atoms so that they are all pointing in one direction, and the end of the nail toward which all of the north poles are pointing will be the north pole of the magnet, and the other end will be the south pole.

This adds another point to the analogy. There is potential life in space before cosmic manifestation takes place, but it takes the vital drive, spoken of before, to awaken it to active life. Acting in this manner, this vital drive is called Fohat.

We may cite experiments to illustrate the teachings of Theosophy, but we must not think that Theosophists have built up a system of reasoning based upon the results of experiments. For instance, we do not observe the behavior of an electric current around a rod of iron, and make the deduction that therefore cosmic life behaves in like manner. This would be approaching the problem in the wrong manner. A Theosophist observes experiments, and sees where they corroborate the Ancient Wisdom, which teaches that the laws of nature, as for instance, the mathematical laws we have just been discussing, are derived from the very life (the automatic will, you may remember) of the Divine Being embodying itself as this or any other universe.

All of the foregoing has been by way of laying the foundation for the setting forth of one of the most wonderful of all the doctrines of the Ancient Wisdom. This concerns the inner and vital and spiritual structure of the planets and stars. It is well that the readers should have a thorough understanding of this foundation before proceeding farther; for they will assuredly marvel, and will possibly experience some difficulty in comprehending the teachings as set forth in this and succeeding chapters, unless they are familiar with this particular mode of thought.

Consider then once more the paradigm we used to depict the nature of an entity.

Energy	Matter
7	1
6	2
5	3
4	4
3	5
2	6
1	7

Here we have again the vertical line representing the stream of consciousness descending through seven levels or planes of space, and building for itself on each plane, bipolar vehicles, making the whole a bipolar manifestation of wide ranges of experience. The student is emphatically warned against thinking of these levels, or planes of consciousness, as placed one above the other like layers in a cake. The methods used here for portraying them are for convenience only, and are not to be taken as a true picture. It is much like thinking of radio broadcasting stations as they come in on the dial, with the stations operating at a high frequency at one end, and those of a low frequency at the other. Everyone knows that these stations are scattered over the country, and that their waves interpenetrate each other all the time. Passing through your room at this moment, are the waves from all the stations that are on the air, and many other vibratory impulses that are not originated by any broadcasting station; yet none of these waves interferes with the others, and we have only to turn the dial in order to select a station. Thus a little thought will convince us that all the planes of consciousness, like all rates of vibration, are everywhere, and that were it possible to "tune in" on any other than that in which we now function, we could do it from our armchair.

This paradigm is then a convenient manner of portraying the constitution of any entity. Since it follows that wherever we see a physical form, it must have its energic counterpart, we learn to view all nature as alive. A planet is alive, and it is to the end that we may understand something of its inner constitution that these pages are written.

You will recollect that we have dealt with the passage of the life-atoms from one plane of consciousness to another, and we stated that this is a process common to all things. And since all things are alive, and are compounded of lesser entities, these lesser lives are likewise continually passing from plane to plane, or from element to element in the inner and outer constitution of their "Lords" — i.e., the greater beings of which they form a part. But just what is it that makes the "Lord" the being that we know as a man, animal, tree, atom, planet, or sun? It is simply the degree to which it has brought out in the process of evolution (the fifth jewel of wisdom) the hitherto latent potentialities and qualities that will ultimately make of it a god, and then a still higher god.

All the forms of life we have spoken of are stages which are necessary for an entity to pass through as it treads the path of immortality to ever greater self-development. Thus an atom is an atom by virtue of its atomic qualities, which require an atom for their expression. The plant is a plant by virtue of its greater development in this hierarchy, so that the still lower forms of life are no longer suited to represent its more awakened faculties. The case is similar for any class of being. A human is a man or woman, not because of the shape of his or her body, but because the body and brain are just the kind of instruments required to represent what a human is, in the relatively awakened state of once-latent energies. Similarly, only a planetary consciousness could manifest as a planet; only a solar consciousness could manifest as a sun, and only a universal consciousness could manifest as a universe.

Let us now go back in thought to the birth of the planet we live on. This planet came into being because there was the need for it, and that need was felt by the planetary consciousness as the urge to become. It could become a world, because it had been a world before, perhaps many times, though a less developed world, learning the lessons that worldhood could teach it; so that being a world was becoming a habit. During its rest periods between worldhoods, its core of pure consciousness had been drawn into its highest essence, in which state it had rested for billions of years. Its one-time sevenfold manifestation had long since ceased to function as the carrier of the life-energies of that planet, and had been dissolved away, somewhat as an army is dismissed after a war. But, just as the able-bodied men of a nation are under natural contract to join the army in future conflicts, so these hosts of subservient entities, having enjoyed individual experiences for billions of years, were under contract to respond to the call to reunite and form a new embodiment of the one-time planet.

You will quite probably have observed that many things are sevenfold, and so you will not be surprised to learn that there are seven main classes of these "subservient" entities, each of these main classes with numberless divisions built upon the number seven and its various powers. Just what these main classes of subservient entities are will be explained later, though we might say now that humanity is one of them.

Now all of these entities had long ago left the physical plane and at the time of the reawakening of the cosmic ray of Consciousness were upon the highest plane possible for them, that represented by the top line in the paradigm (energy 7, matter 1), and were there to meet the "descending" ray. As this ray "descended," its subordinate entities followed it through plane after plane until the lowest of them was reached. But —

We explained how lightning was the result of a built-up potential between two clouds, or between a cloud and the earth, which breaks through when the stress becomes too great. It is just where the resistance is weakest that the lightning flash occurs. We might think of the transference of entities from one plane to another as the result of the building up of a life-potential until there is a breaking through — at the weakest point, shall we say? That might not be putting it very well, but we should explain that there are what in Theosophical parlance are called "laya centers," or dissolving points. The word *laya* comes from the Sanskrit root *li*, to dissolve. These are natural gateways through which the entities may pass from plane to plane. A seed, for instance, is such a laya center, for through it flow streams of energy, carrying with them hosts of life-atoms which clothe themselves in material atoms, which go to make up the plant; and the process is similar in the case of an egg, which results in a bird or an insect, and so on.*

*It is interesting in this connection to note that the solar system was referred to by the ancients as the Egg of Brahma, and the earth as the Mundane Egg.

Now these entities, these life-atoms of various grades, stream through such a laya center on the highest cosmic plane in the constitution of a planet to be, and collect around the laya center in ever greater numbers, until their number becomes so great that they form a globe or world on that high plane, this being the first step towards the manifestation of that planet-consciousness as a world. The consciousness continues to project itself, or rather to send its ray, deeper into matter, and swarms of life-atoms descend with it, entering a second laya center around which they gather, and build a second globe, the second step in the embodiment of the planet consciousness. Thus we see that these globes are actually built by and of these life-atoms themselves.

This process continues until the consciousness has reached and quickened the seventh and lowest of the cosmic planes. And then a strange thing happens. The entities attain a momentum by their passage through the seven planes, and commence a reascent through them, since they cannot go any "lower." This impetus is the result of two lines of evolution; their own, and that of the entity embodying as the earth. Furthermore, in retracing their steps through the higher planes they cannot run against the tide of entities coming down through the laya centers (for there are entities coming down all the time), so they must find a new laya center for their upward journey, and build around it a new globe, because the passage through any plane, whether upward or downward, takes millions of years. Thus we find that there are two globes built of entities on each of the planes of consciousness except the lowest and the highest planes, on each of which there is but one globe. When the entities have completed the round trip from the highest plane to the lowest and back again, they have built twelve globes, represented here by circles.

Spirit								Matter
7				O				1
6			O		O			2
5		O				O		3
4	O						O	4
3		O				O		5
2			O		O			6
1				O				7

The passage of the entities downward through all of the planes and back again is referred to in Theosophical terminology as a Round. In the Round just described, the entities were employed in building these twelve globes. When there was a general exodus from one plane for the purpose of building a globe on the next, whether going up or down, a certain proportion of the entities of all the seven classes remained behind on each globe "to hold the fort," in order to preserve the globe intact that it might receive the hosts of entities on their return journey during the succeeding Round. An analogy that might be helpful in this regard, but which cannot be pushed too far, might be stated thus: Suppose a man had twelve empty beehives that he wished to have filled with swarms. He obtains a swarm and "plants" it in the first hive, where they are contented for a time; then part of them leave to form a new swarm and are put into the second hive. From there later on a swarm issues forth to settle in the third hive and so on until all of the twelve hives are supplied with bees. Each time a certain portion of the swarm is left in the old hive. While this is a clumsy analogy, it may serve to illustrate how the hosts of entities can desert a globe, and yet leave a certain number behind to keep the globe a living thing. It is indeed an economical system, for it does away with the need of rebuilding the globes in any subsequent Round. This planetary constitution functioning as twelve globes on seven planes of consciousness is termed a *planetary chain*. Thus we see that each of the globes is a focal point through which are streaming hosts of lives at all times.

Every globe we see in the sky, all of the stars, including our visible planets, are sevenfold bipolarities with their own twelve globes, one only of which we can see, the one that happens to coincide with the level on which we are functioning.

The Ancient Wisdom tells us that there are innumerable worlds (planetary chains), none of whose globes appear on our physical plane. These constitute the invisible planets, which are none the less real, even though our limited senses fail to bring us report of their existence. It should be further remembered that the sun, and consequently all the stars, are sevenfold, forming solar chains as contrasted with planetary chains.

Doubtless you are wondering on which of the twelve globes of our planetary chain we happen to be. So I hasten to lay before you the unwelcome fact that we are on the lowest globe of all. All of the other globes are quite invisible to us, but we shall see them, as, in the ages-long course of evolution carrying us along, we shall pass from one to another. From this it follows that we see only the lowest globe of the visible planetary chains in our solar system, and for the most part, the lowest of the globes of the stellar chains. What a glorious sight it would be if we were able to see all the globes of all the planetary chains we call Mercury, Venus, Earth, Mars, Jupiter, Saturn, Uranus, Neptune, and Pluto, and how much more glorious would be the sight of all the globes of our sun, could we see them! But we are assured that these things can be seen by those who have the faculty of inner vision, and this faculty is the reward of discipline and training.

It is interesting in this connection to learn that modern science has now proof of the existence of invisible worlds in the supernovae that have been recently discovered. For instance, there is the companion star of Epsilon Aurigae, a Red Giant star. This companion is a giant star, so large that it has been calculated that the sun and all the planets on their orbits as far out as Uranus could be comfortably accommodated within it! Furthermore, when this star eclipses its smaller companion, the latter does not diminish appreciably in brightness — which it would do if an opaque or even semiopaque body should come between it and ourselves. At the time of this writing (1940), though attempts have been made to photograph it with infrared sensitive films, we understand that the star has not been photographed, though the attempt to do so has brought to the attention of astronomers the existence of other stars hitherto quite invisible, and therefore unsuspected. It has since been found that the companion star emits strong X-rays, and only a "black hole" could do that. It is findings such as these that corroborate H. P. Blavatsky's teachings.

IX

The Kingdoms of Nature

In the preceding pages we have endeavored to set forth the teachings of the Ancient Wisdom regarding the true nature of things. While we have particularized on the construction of a planet, it should be understood that the paradigms used for that purpose are descriptive of the makeup of any entity whatsoever, giving assurance of the truth of the teachings. For what is true of the great is also true of the small: fundamental truths are universal. Now we are to go a step farther and consider the distinctions between the various classes of beings that inhabit our earth, and so go on to a study of the nature of the lives which have brought forth other worlds, and their relations to ourselves.

First then, we are prompted to remark on the supposedly scientific fact that the earth itself is composed of what is popularly termed the entirely inanimate mineral kingdom, supposedly built by the processes of heat, pressure, concretion, and sedimentation. It is inhabited by the vegetable, animal, and human kingdoms, of which the first is said to be barely animate, the second animate beyond a doubt, but not so highly developed in respect to brain power as the third, the human stock. It is further popularly supposed that in the human family there is no differentiation into species as regards physical forms; that among the animals there are so many differences that there are almost no resemblances among some of the species, as for instance, an elephant, a serpent, and a fish; again, in the vegetable kingdom, that there are a great many varieties, though perhaps their differences are not so marked. However, what really establishes the fact that the representatives of these three kingdoms are animate is that they *grow*.

There is little of spiritual or even intellectual value to be gained from a study of the earth aside from a tabulation of facts, unless we apply the principles of Theosophy as taught in the doctrine of evolution. Otherwise we can give no satisfactory reasons for the diversified forms of life in any kingdom, let alone the existence of the various kingdoms themselves, obviously so different yet bearing fundamental resemblances to one another, as shown by the study of cytology.

First, then, let us consider that a kingdom of nature is like a grade in a school wherein a pupil studies the lessons appropriate to that grade. No children enter a school grade until they have proven by examinations and test questions successfully passed and answered that they have learned the lessons of the last year's grade. This analogy is particularly apt when we remember that the understanding is already within, and that education, as Socrates demonstrated, is the process of bringing forth that which is within. Thus, Theosophically, the kingdoms of nature are looked upon as schools of experience for the entities composing them; and only after ages of experience in any one grade is the life-wave or tide of entities composing that grade or kingdom ready to pass on and form a higher kingdom. It is once more the old, old story of the bringing forth of latent characteristics and potentialities heretofore locked up within the entities themselves, in other words, evolution.

Let us abandon the wornout idea that evolution is a change of form to suit the change of environment. The forms change, truly, but chiefly so that they may the better serve to house the ever-growing entity. What then places an entity in any particular kingdom? The form it wears? Assuredly not, for the form merely stands as a symbol of what the inner being really is. That which places an entity in any particular kingdom is the degree to which it has evolved its hitherto latent characteristics, so that it can build for itself a form which identifies it as being either a plant, an animal, or a human

Nature is constructed of hierarchies throughout, each of which contains the possibilities for many degrees or stages of growth. The lowest stage for any hierarchy is *its* atomic kingdom. Thus it is taught that while, generally speaking, our atoms represent the lowest grade in our particular hierarchy, there are atoms nevertheless that are far higher than anything we know of in our hierarchy, because they are the building blocks of hierarchies vastly superior to our own. Certain stages above the atoms are the molecular, the crystal, the living cell — this last being a stage below the animal and vegetable, though it is a part of both and partakes of the characteristics of both.

With these thoughts in mind we must remember that an atom in this hierarchy is an atom because as yet it is an atomic consciousness and is at present incapable of building for itself anything but an atomic body, complex as that is now known to be. A plant entity or monad long ago evolved out of the mineral, and previously from the atomic and molecular stages, and has developed its inner energies and powers to such an extent that nothing less than a plant body will suffice to give the plant forces their scope for action. But even its relatively greater degree of evolution is not sufficient to enable it to build for itself the body of a human. Indeed, there are aeons before it in which it must learn all the lessons of plant life and of animal life before it can undertake to learn the greater lessons of humanity.

Following along this line of thought, we make two significant deductions; first, that it is not the form which identifies an entity as belonging to any particular kingdom, but the entity places itself there by being what it is, and builds for itself a suitable form which identifies it. Second, all beings must pass through all conditions before the goal (for this particular life cycle) is reached. Thus all entities were once, are now, or in some future age will be in the human stage of evolution, not necessarily having our present shape, but forming nevertheless the human kingdom. This is true for all entities. Thus there is no fundamental difference between the essence of the consciousness of a dog and that of a man. That is why brotherhood is universal and must be recognized as taking in all things. The differences are simply due to relative progressions along the pathway of evolution.

Science says that all below the animal and vegetable kingdoms comprises the mineral kingdom, but for many purposes Theosophists find this too broad a statement. They point out that because it is the evolutionary standing of the beings themselves which determines the kingdoms to which they belong, and not the materials of which their bodies are composed, these materials in themselves do not belong to the customarily recognized four kingdoms, but are parts of other kingdoms hitherto unrecognized by science. These are the elemental kingdoms.

Among the most interesting and at the same time most difficult technically of our teachings is that concerning the life-atoms and the part they play in building man's nature in all its reaches of consciousness-substance. The subject, moreover, is made difficult because we tend to draw a distinction between the physical atoms and life-atoms, and this tends to confusion.

Life-atoms exist on all levels of the constitution of both universe and humanity. They are evolving, as all things are, and, passing through our sphere, they are known as the "elementals." Even these are arranged by nature into seven classes, four of which relate to important elements of the earth's own structure.

In more advanced theosophical studies, illuminating teachings are given about the seven classes of life-atoms or element-principles in universal nature, sometimes called the *tattvas*, but more than a brief account of these would go beyond the scope of this book. This much, however, may be said: just as we are accustomed to study the constitution of man in terms of the higher triad and the lower quaternary, an explanation of the tattvas may be presented in a similar manner. In what follows the words for the higher triad are given in Sanskrit, while for the lower quaternary we use the more familiar English words. Thus:

Adi-tattva	Primordial or Divine
Anupadaka-tattva	Literally, "parentless," coming into being *sui generis* from Adi
Akasa	Spirit
Fire	Elementals associated with it known as salamanders
Air	Associated elementals, sylphs
Water	Associated elementals, undines
Earth	Associated elementals, gnomes

It is probable that fairy tales and folklore are the result of half-forgotten memories of some aspects of the Ancient Wisdom and tell us of many instances of people with second sight. It should be made clear, however, that Theosophy does not concern itself with these.

When the "three elemental kingdoms" are referred to, this is but another way of viewing the subject, much like stating that humans are composed of spirit, soul, and body. And this correlates well with the threefold division of the cosmos as *gods, monads,* and *atoms.*

Before leaving this chapter, a few words on the degrees of consciousness attained by various classes of entities may be timely. Of all the innumerable degrees of consciousness, we know the human only, and can to a certain extent judge the consciousness of an animal. In the plant world, the condition of life is past our ken, just as it is in beings higher than the human. So much is this the case when we consider the atomic lives about us that many declare we have no evidence at all to show the atoms are alive, and that it is but a side-stepping to say that all things are alive, but in differing manners. In the following paradigm an attempt is made to show the relationships between the various states of consciousness enjoyed by the entities on the several rungs of the ladder of life.

In the first place, we establish the bipolarity light and life. Light, as the upper pole of life, is the divine consciousness from which life, as a more or less automatic operation, is derived.

In the context of the following paradigm, Light is construed as Divine Consciousness, from which spring the numberless forms of living beings, for which the general term "life" is given.

.... Light ....
...... HUMAN .... Light ....
.... ANIMAL .... Light ....
.... PLANT .... Light ....
.... Life .... MINERAL .... Light ....
.... Life .... ATOM ....
.... Life ....
.... Life ....
.... Life ....

Observe that life for the human hierarchy is on the same plane of perception as light in the atomic hierarchy. Consequently, we seem to be aware of diametrically opposite portions of two entirely different hierarchies: in the case of our own, it is the lower pole, life, and in the case of the atomic, it is the higher pole, light. Thus, it may be that the life of the atom is the lower pole of the light that is visible to us, and that is consciousness itself on another plane.

X

The March of the Monads

Suppose it were possible to make an attempt to remove a man from the universe. What would happen to him? Where could he go? If the word "universe" is all that the name implies, we see at once that there is no such thing as a place where the universe is not. So in taking a man out of the universe there would be no place for him at all. He could not even wander about like a lost soul, for the simple act of wandering is a change of location which would be impossible for him. Nor would he be of any size at all, for he would not be able to take up space, there being no space for him to occupy. Furthermore, the attempt to remove him from the cosmos would be fraught with certain difficulties. In leaving the universe, the man could not take with him anything belonging to the universe. That is, he would have to return to the various planes of consciousness, the materials and substances of all sorts which he derived from them for the purpose of building for himself his various sheaths or vehicles of consciousness, which we have learned to call his "principles." Upon careful study, we find that no man owns anything of himself, but every atom of his being is derived from the common parent of all things.

Therefore, taking him outside of the universe would not only put him nowhere, but would reduce him to nothing at all.

Now, one is faced with certain difficulties in elaborating a theme such as ours, which is the building of the world. Not the least of these difficulties is the fact that the beginning of the world is no beginning at all, but a continuation of the eternal drama of life. So true is this that, had it not been for previous aeons of experience, the hosts of entities that compose our world would never have been brought to that condition in which it was possible for them to build this earth some four billions of years ago. And such experience was none other than the building of other earlier and less evolved worlds; for practice makes perfect. We have visible evidence of the last embodiment of the Earth. Long before the Earth came into being, the moon was a living planet, and its planetary consciousness was the same which, after leaving its outworn shell, reembodied itself in due course as our Earth. The moon is the last vestige of the previous embodiment of our earth entity.

We must remember that in principle the building of a world is essentially the same as the building of the constitution of any entity, whether an atom or a galaxy. Whatever it may be, it is a mystery indeed, for in every case, the eternal pilgrim is clothed from time to time in living substances composed of lesser entities, his own life-atom children. His entire constitution, both inner and outer, is built in this manner, because it has been so compounded hundreds of times before, and the same entities, the countless subservient lives which in their aggregate form his various "bodies," are karmically linked with his destiny, and are the same that have built his vehicles in previous lives. But even as he is the indwelling spirit, and grows and evolves while embodied, so do these subservient lives learn and grow as the result of their experience in making the vehicle for their "god." Slowly, as the ages roll on, these subservient entities learn how to make better vehicles, more adequately to enshrine the indwelling divinity. So evolution attains its end by a process of helping and sharing. In this way it is an easy matter to understand why it was stated that the beginning of the world was no beginning at all. All the entities subservient to that Cosmic Being which embodied as the Earth Chain (Chapter VIII) had served that divinity for uncounted ages, and had learned by experience how to make a better world in which the divinity might clothe itself.

Now, what are these unnumbered hosts of entities, if not the seven kingdoms spoken of in the preceding chapter? Indeed, here is the unfolding of the mystery: The seven kingdoms of nature, the three elemental kingdoms, the mineral, the vegetable, the animal and the human, in their vast aggregate are the body of the Earth, and the building of the globes of the Earth Chain, as explained, was brought about by the passage of the members of those seven kingdoms from plane to plane.

In order to simplify the picture which we have now to paint of the building of the Earth Chain by these seven classes of entities, we are going to suppose that there are globes to be built upon the four lower of the cosmic planes only, as in the diagram.

Spirit			Matter
7	O		1
6	O	O	2
5	O	O	3
4	A O	O G	4
3	B O	O F	5
2	C O	O E	6
1	D O		7

This makes seven globes instead of twelve, though we have learned that twelve is the complete number of globes which occupy all seven of the cosmic planes. This will bring out the dissimilarity between the nature of the globes built upon the lower planes, which are referred to as the seven manifested globes, on the one hand, and the five unmanifested globes which occupy the three higher cosmic planes of the earth's constitution, on the other. When we speak of manifested globes, we do not necessarily imply that they can be seen with the physical eye; indeed, all but one are invisible, since they are composed of finer grades of spirit-matter than we can see; and again, when we speak of unmanifested globes, we do not mean that they do not exist, but that their nature is subjective, as contrasted with that of the objective globes. Subjective though these higher globes may be, they are entirely real for those entities that inhabit them; yet their nature is such that language is not adequate to describe them. These seven manifested globes, are lettered for convenience, from A to G, as shown in the diagram.

Here, then, is the order in which these seven classes of beings marched along the downward arc from the highest spiritual plane to the lowest, or physical plane, and then proceeded along the ascending arc to return to the lofty spiritual plane from which they proceeded. The first elemental kingdom led the procession — supervised by the Master Builders, highly evolved entities that had gained their experience on the Moon Chain — and formed the nucleus of Globe A, and when its work was finished, moved on, giving place to the second of the elemental kingdoms, which thereupon took up the work. Meanwhile, the first elemental kingdom had passed through a laya center (Chapter VIII), and was forming the nucleus of the second globe, Globe B. Having completed its work there, it made way once more for the second elemental kingdom that it might likewise work on Globe B; and so we find that the first of these elemental kingdoms is next building the foundation of Globe C. During all this the third of the elemental kingdoms has made its advent on Globe A.

With the turning of the cycles, these three elemental kingdoms move on, and the mineral kingdom enters the arena on Globe A, which then undergoes a further degree of solidification. The fact is, the mineral kingdom of that time bore no apparent resemblance whatever to the rocks that form the crust of our present Globe D. How difficult it is to explain these things which are so foreign to the modern ideas of the building of the world! One can almost hear the cry of indignation from those who fail to accept the idea that the earth is in reality embodied consciousness. A little patience will repay us well, and in the course of time we shall see where the geological facts as read in the rocks and strata fit in with the scheme as enunciated by Theosophy.

Our picture, then, as far as we have painted it, shows Globe A as having been built to a certain extent by the three elemental kingdoms, and the mineral kingdom; Globe B as having been built to a lesser extent by the three elemental kingdoms; Globe C by the first two elemental kingdoms only, and Globe D in the process of having its foundation laid by the first of the elemental kingdoms.

Think always of the passage of the kingdoms from one globe to another as the marching of armies of life-atoms and archetypal entities from one plane to another, as before described. When the work of the first elemental kingdom was completed on Globe D, it passed on up the luminous or upward arc to the plane on which it had previously functioned when it built the nucleus of Globe C, building now on that same plane a second globe, Globe E. Meanwhile, all of the three other kingdoms move on a step, the second elemental kingdom to Globe D, the third elemental kingdom to Globe C, the mineral kingdom to

Globe B; and the vegetable kingdom now makes its first appearance on Globe A. This process continues, with all of the kingdoms making the grade step by step, so that the animal kingdom appears on Globe A, and then the human kingdom is the principal actor on Globe A, moving on to Globe B, then C, and so on. By the time the human kingdom makes its first appearance, the first of the elemental kingdoms is about to build its seventh globe. We left it on Globe E, whence it passed to Globe F (on the same plane on which it had previously built Globe B), and it has now reached the plane of Globe A. On that plane it builds the highest of the seven manifested globes, Globe G.

Having done its work, the first elemental kingdom goes to sleep, giving place to the second elemental kingdom on Globe G, after which this second elemental kingdom goes to sleep in its turn giving place to the third elemental kingdom. In turn the mineral, vegetable, animal, and human kingdoms finally reach and complete their work of building Globe G, after which they fall asleep in their turn. This "sleep" we call obscuration or dormancy. It is a temporary state of rest. The purpose of the journeying of these seven classes of monads from plane to plane is to build the globes of the Earth Chain, in obedience to the need for the earth entity to embody itself at the beginning of its new cycle of activity. And now the globes have been built, and the globe chain exists as a complete entity. The going to sleep of the seven classes of entities after they have built these globes in no sense meant that their work was accomplished, or that the life of the earth had drawn to a close. But just as the heart beats and rests, so the life of a planet pulses through cycles of activity and rest. We have a technical term for the completion of the life-waves' journey through the globes and planes of consciousness. When the downward and upward arcs are described, it is said that they have made a Round of the Earth Chain.

We have described the first Round. Actually in the life of the Earth Chain, as in the lives of all the planetary chains (for all planetary chains, like our own Earth, are composed of seven globes), there are seven Rounds. That is to say, the entities make seven complete journeys, the first of which is primarily to build the globes — which is another way of saying, for the embodiment of the earth entity — and the remaining six Rounds are for the pursuance of the life cycle of the earth entity, and for the added experience of the life-atoms or other lesser entities which form the vehicle of manifestation of the earth entity. Actually, we see two lines of growth going on simultaneously, that of the earth entity in its own reembodiment, and that of the hosts of subordinate entities which gain experience on these various planes of consciousness and substance, in their task of building and inhabiting these globes.

Before passing on to the next chapter, one or two points should be brought out, as they are important. In the first place you will remember that we have stated a number of times that the Earth Chain reembodies because it has been a planetary chain before, and will be so again many times in the future. Now some of the most evolved of the entities which had helped to build the Moon Chain, the previous embodiment of the Earth Chain, and which had attained to a higher state of life than humanhood, were actually the first on the scene at the time of the building of the Earth Chain. These exalted beings did not have a direct hand in the formation of the globes, as did the three elemental and other kingdoms, but guided the proceedings, giving a spiritual impetus, if you like, to the process. This was not elaborated before, for there was danger of unduly complicating the explanation. But it is important to understand at all times that the earth is watched over by spiritual beings, who, although unseen, exert a powerful influence upon ourselves and upon all nature as well.

It might be useful here to make a new diagram which will illustrate the correspondence between the relations that the twelve globes of the Earth Chain bear to one another, and the relations that the life waves bear to one another. The Theosophist recognizes at least ten classes of life waves, or kingdoms of nature, and with further study one learns that there are actually twelve kingdoms. The intricacies of the numbers 10 and 12 are apt to confuse students unless they remember that these truths can be given out only in small degree at first, and expanded into more complete teaching as the students progress in their study.

The reader will readily understand that the diagram on the next page is intended merely to show the close correspondence between the globes of the chain and the entities that build and in their great aggregate *are* these globes.

Think of these kingdoms of nature as you do of grades in a school. Pupils advance from grade to grade as they learn the lessons and pass the examinations. Thus, one grade does not become the next higher, but the pupils advance, and the grade then offers its lessons to the younger pupils who take the places of those who have gone on. Whatever advancement is made by the grade itself is due to the growing experience of the teachers and the adopting of new and improved methods of instruction, so that children in the seventh grade, for instance, learn more today than they did a generation ago, but it is still the seventh grade. In like manner, think of the entities as passing through the kingdoms for experience. Essentially they belong to no one particular kingdom any more than to another, but for the time being they are members of one or another in order to learn the lessons of life and experience that evolution has in store for them.

In the case of the work of advanced beings higher than

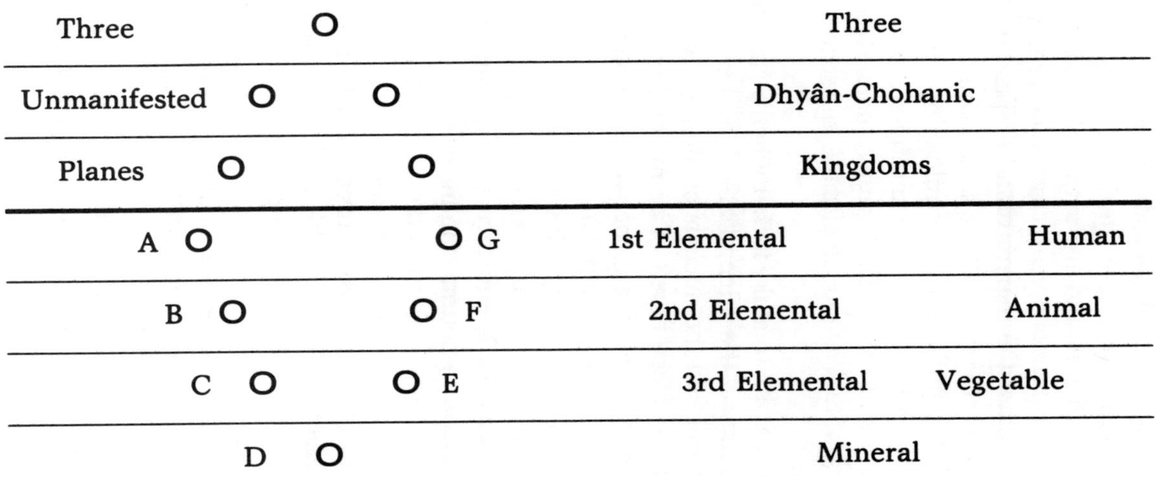

Three O Three

Unmanifested O O Dhyân-Chohanic

Planes O O Kingdoms

A O O G 1st Elemental Human

B O O F 2nd Elemental Animal

C O O E 3rd Elemental Vegetable

D O Mineral

the human who presided over the building of the Earth Chain at the beginning of the first Round, this work may be likened to the founding of the school used above as our analogy. A school has its beginnings, not with babes and children who make up the classes, but with the architects, builders, and instructors, all of whom had been to school in previous years, and who are now able to gather together the materials and prepare the buildings for the incoming pupils. The school consists first of the grounds, the buildings, the equipment (which must be gathered together under the careful supervision of the founders), and then, of the pupils themselves, in their various grades. Without all of these, the school could not function.

Another point to make is that when a class of entities, as for instance, the mineral kingdom, leaves one globe in order to build the nucleus of the next, we are not to think of one globe bereft of its mineral kingdom, for if such were the case the globe would have derived little benefit from the passage of the kingdom through it. Actually, it is the life impulse which leaves the globe, and passes on, causing the subsequent growth of the kingdom on another globe. The material or substance portions of that kingdom, of course, cannot leave the globe where it has been built, but it sinks into a state of relative inactivity, retaining a certain amount of the life impulse, but very little indeed as compared with the inrush of vitality which had once quickened it. The residue of these kingdoms left on each of the globes remains there, waiting for the time, aeons later, when the life impulses, which had brought about the establishing of those kingdoms on the globes, return to them in the course of succeeding rounds, and reawaken them to renewed activity. There are many mysteries connected with this teaching which can be unfolded only in part, and in due course.

XI

The Earth's Former Self

The outline presented in the preceding chapter as to the manner of passage of the seven classes of entities through the four lower planes, building in their progress seven globes which function together as the Earth Chain, was of necessity brief and incomplete. It is impossible to do more than touch the high spots of so complex a teaching. In this chapter we shall endeavor to clarify the picture by pointing to a period prior to that of the formation of our present chain, and study at some length the causes which led to the building of the one with which we are so closely linked in our karmic destiny.

First, we have stated that the Earth Chain is the reembodiment of a former planetary chain. Is there evidence of such a former planetary chain having existed? There is, and we must call the moon to the witness stand. The moon, which exerts so strange an influence upon the earth, is the remains of the last embodiment of the earth entity. The case is much like that of an Egyptian who perhaps died some thousands of years ago, and who reincarnated and then beheld the mummified remains of his previous body, finding it either in a tomb, or in some museum. He, of course, could not know that this mummy represented his former birth, and only one gifted with inner vision could trace the thread of destiny and show him how he had been drawn to the side of this relic of his own past, his "moon."

In like manner we have been taught by those who have the means of knowing, that the moon is the visible reminder of our own past. Aeons ago, when the earth entity was embodied in what we call the moon, all of the hosts, armies, legions of entities that now form this Earth, had built and inhabited the then living moon. It supported its own races of intelligent beings, human we might call them, because occupying the same rung on the ladder of evolution that we now hold, though it does not follow that in appearance they would be recognizable as humans.

When the facts are known, we shall find that during certain periods of human development on this our own Earth, we took on appearances that bore little or no resemblance to that which we now know as the human form. However — and here lies the burden of this present chapter — the present form of the human body (to speak of man's inner vehicle is premature) is the direct outgrowth of the forms which it had in the earlier periods of man's development on Earth. Had the urge for self-expression taken a different course in those earlier times, our present bodies would not be what they are today; and, extending this teaching farther, had it not been for the various forms and changes that human evolution brought about on the moon in its heyday, the human family would not have become what it is. Even though some of the then humans have passed out of the human stage, they all helped to lay down the pattern around which the present human race is molded.

To sum up, then, we are evolutionally at the present time the result of all we have been through ages past, not only on this Earth, but during the cycle of its previous embodiment as the moon, and indeed, still farther back, through still earlier embodiments, as far back as you care to go. Remember that part of us belongs to the Earth, and that that part of us is of the hierarchy of life-atom children which the Cosmic Being brought forth from its own essence aeons ago. Thus humanity follows the earth entity through its many reembodiments as globe chains, and helps to form those globes. As said before, practice makes perfect; and who would say that we have yet made a perfect world?

Just as we pointed out that the present human form owes its character to the evolutionary urge of aeons ago, we now declare that the same is true of the bodies of the animals and the plants, with this difference, however: man himself is largely responsible for the shapes in which these lower entities clothe themselves. Herein lies one reason for our stressing, as we did, the difference between the inner monad of any creature and its outer carapace. To be sure, we have said time and again that each entity builds for itself just the kind of body that it requires for its own degree of evolutionary development; yet we now have to postulate the teaching that these bodies are woven around certain patterns. In large measure these patterns were furnished by the human kingdom of aeons ago. To be more explicit, the humanity of the Moon Chain set the pace which is now followed more or less, but with continual improvement, by the entities of this, our Earth Chain. The thoughts and emotions of this lunar humanity shaped themselves into the patterns around which the "lower kingdoms" of this present Earth have built their bodies, just as we of the present time are fashioning by our thoughts (good as well as evil) the astral forms around which will be shaped the entities of the lower kingdoms which will some day inhabit that future planetary chain, the reembodiment of our Earth.

Thus it is stated that the beautiful and shapely creatures of Nature, both in the animal and the vegetable kingdoms, built themselves around the patterns furnished by the "grown-up" thoughts, shall we say, of a high and lofty type emanating from that humanity of long ago. Similarly, the base and ignoble thoughts and unconquered desires have resulted in the formation of the patterns for the venomous and baleful creatures. The potency of thought and the seriousness of our responsibility are brought home to us when we reflect on the law of reembodiment, and that this Earth too shall one day reembody, and that it shall be peopled anew with creatures of all kinds; and furthermore, that we of today are setting the pace in large measure which we and the kingdoms below us shall follow in the aeons of the far distant future; yet, not so far distant, but very close to us in one sense, for we ourselves shall take part in such stirring times again, in the building of a planet.

Taking up once more the theme of the first Round, as it was called in Chapter X, we see that when the entities of various kingdoms passed through the laya centers on the various planes of consciousness, and fashioned therein the seven globes, there were plans definitely outlined for them to work upon. The elemental kingdoms came first, to provide the materials, just as the potter requires clay before he can work; and then the higher creatures entered the arena, and fashioned, out of the materials at hand, the bodies they required, building them around the patterns that had been impressed, shall we say, upon the astral light. The astral light, we are taught, may be likened to a great photographic substance or medium which records all events, thoughts, and deeds. The seer can read the records of the past as they are recorded on the astral light.

All are familiar with the story of the Flood common to all mythologies. Noah's Ark represents the astral light, carrying within it one pair of every type of creature that shall live upon the earth. During the flood, which for the purposes of this study represents the long period of oblivion between the life of the moon and its reembodiment, the Earth, these creatures are carried across the waters of space until the flood subsides once more, and they alight upon Mt. Ararat. The animals in the Ark (and Mr. and Mrs. Noah for that matter) are the patterns around which the entities of the chain are to fashion themselves. Mt. Ararat is the Earth as built by the elemental kingdoms and made ready for the advent of the kingdoms to follow.

In a succeeding chapter we shall endeavor to set forth some of the teachings concerning the cycles of development of the races of entities during the rounds of the Earth Chain, which, we shall understand, follow the general type to which all planets adhere; for nature is governed by one law of life everlasting.

XII

Rounds and Races

When we presented the paradigm of the seven planes of consciousness with twelve globes functioning therein, we were using a simple diagram to illustrate a very complex teaching. Now there was no more point in showing the planes as one above another than there would have been in drawing the lines vertically, except perhaps that in the manner shown, the truth may be presented a little more readily to the mind unaccustomed to these thoughts. The fact of the matter is that these planes of consciousness are in no wise separated from one another in space or position, and are not to be thought of as layers in a cake. Reflect once more on the analogy of the various wavelengths employed in radio broadcasting. These interpenetrate one another, though, to be sure, the stations are situated at various points all over the world, and the stations come in, moreover, in a certain order on the dial of the receiver. We often think of them in this order, with the stations of high frequency at one end of the dial, and those of a low frequency at the other; but you will readily understand that while we can tune in one station to the exclusion of all others, all the wavelengths are passing through the room all the time, and all at once. Similarly, this matter of the planes of consciousness, difficult to grasp as it is, may be brought home to us if we realize that all a paradigm can do is to present a symbol or pattern which shows the *modus operandi*.

Having carried these studies thus far, we are more familiar with the idea of the entities pursuing their journeyings through the planes of consciousness on their first passage around the globes, building the seven manifested globes of the Earth Chain, and on succeeding Rounds returning to those globes for the further experience of the life-atoms and of the earth entity likewise. We recollect further that in doing so they are following a pattern which had been outlined aeons ago, when the former embodiment of the Earth (the moon) was in its heyday. We now propose to set forth some of the doctrines concerning the cycles of development of these hosts of entities here on earth.

In the first place the work of evolution is an age-long process. Actually, it is roughly two billions of years since the first elemental monads commenced to build Globe A. The question is further complicated when we learn that the various kingdoms do not run the course in the same time periods. While our neat picture of the monads following one another — each one leaving a globe when its work was accomplished, making way for the next class, which in moving on makes way in its turn for the class to follow — is correct enough in outlining the general plan, it requires certain emendations. As an instance, it should be mentioned that the grossest kingdoms such as the mineral or vegetable find the ascent along Globes E, F, and G far more difficult than does the human, and their progress through these higher globes will be correspondingly retarded. This gives the human kingdom an opportunity to "catch up" with the lower kingdoms, to a certain extent. With this acceleration on the part of some entities and the retardation on the part of others, the thing works out in the long run, so that at the end of the seven Rounds of the planetary chain, all of the classes of entities finish the race together. So it is easy to understand that at the point in the earth's life at which we now stand (slightly beyond the middle point, as will be explained), the lower kingdoms have not all given way to the advent of the human. We have already partly "caught up" with them. In view of these facts, we shall not attempt to outline the time periods of the lower kingdoms, but shall touch on the racial periods of the human family alone. Nor shall we at this stage attempt to discuss the appearance of the human race during the early period of the earth's history, that being left for a future time.

According to the old Brahmanical tables of the cosmic periods, the length of time elapsing from the early formation of a world to the corresponding point of early formation of its reembodiment is somewhat over eight billions of years. During that time the old world will have lived its life and passed through its death stage, and will be ready to reembody. The period of activity equals the period of rest between embodiments, and so a simple division shows us that the entire life period of a world from the time of its birth to the end of its present life cycle is roughly four billion years.

We have spoken of the seven Rounds which comprise the life cycle of a globe chain. One might suppose that to find the duration of one Round, it is necessary to divide the life period by seven, but this would not take into account certain rest periods between the Rounds which are called "obscurations," in which the Earth seems to lie quiescent (the present condition of certain of the planets of our solar system). Other factors might be brought in, but it will suffice to say that the period of one Round is a tenth of the life period of the globe chain. Thus a Round lasts for about four hundred millions of years.

A Round, as you will recollect, is the passage of the life-streams through the seven globes, yet the time spent on any one globe (taking into account the rest periods between globes, etc.) is about one-tenth the duration of a Round, or forty millions of years. Now while a race of beings is on any one globe, its stay there is marked by certain phases of growth, seven in number, and, in speaking of humanity, these phases are referred to as "Root Races." Allowing again for periods of quiescence between Root Races, we find that the duration of a Root Race is about four millions of years. Taking into account certain factors, to attempt to explain which would be impossible here, each Root Race may be divided into subraces of 432,000 years each. One could subdivide endlessly, and indeed, for deeper study we might later find it advisable to do so; but for our present purposes we have carried the point far enough. To clarify the teaching, let us tabulate these time periods in round numbers, and bear in mind that these figures apply to the human life wave only.*

	Years
Life of the Planet	4,320,000,000
Duration of one Round	432,000,000
Time spent on one Globe	43,200,000
Duration of one Root Race	4,320,000
Duration of one Kaliyuga	432,000

*No attempt is made here to set forth the cycles in detail, and it must be borne in mind that the numbers in this table are round numbers only. The problems which deal with these racial periods are numerous and perplexing, for the Teachers tell us that the time has not yet come when the ultimate keys may be given out.

When we call to mind that some of the geological deposits show that on this earth of ours sedimentation began in this Round between three and four hundreds of millions of years ago (there are some geologists who would place it at a billion years), we must realize that the old earth has witnessed many changes. Geologically speaking, there is no telling for certain what the earth was like before sedimentation began. The various strata that are considered to be the oldest, having been laid down in the Azoic times, actually date back several Rounds! What does that mean? Simply that we, with all the other entities, in making the Rounds of the globe chain have had time to pass around the complete chain several times!

The teaching is that we have made three complete Rounds, and are now in the process of making the fourth. Furthermore, it was mentioned previously that we are at present on the lowest of the cosmic planes of the solar system, which may be expressed in another way by saying that we are on Globe D. Since this is our fourth time around, it is plain to see that we have completed approximately three and a half Rounds. For those interested in knowing exactly where we stand, it might be said that we are nearing the middle of the fourth subrace of the fifth Root Race on Globe D, during our fourth Round. So far as the life of the Earth Chain is concerned, this is known to be just a little more than half spent, or to put it in more familiar terms, this earth is middle-aged at two billions of years!

The lesson that we would bring to mind is that this matter of the Rounds and races is not a faraway speculation, but is a process that is going on all the time. We *are* making the Rounds of the globes of our chain, and in some millions of years hence, we shall find ourselves as the human family functioning on a higher globe, and then still higher, and so on, until we have passed once more completely around, and shall find ourselves yet once more here on Globe D, some 400,000,000 years hence!

XIII

Truths Stranger than Fiction

The threefold classification of man's constitution, body, soul, and spirit (Chapter V), is not an arbitrary tabulation, but rather one of various manners of classifying the energies and powers of this wonderful being we call "man," for there are different methods of approaching a true understanding of what he really is. We shall study him as a sevenfold entity, but a thorough understanding of the threefold division is first requisite.

In the light of what has been set forth in the preceding pages, we are to consider man as composed, not of one entity alone, but of *three distinct entities.* Although *no thing* ever came from nothing, all things had an origin in this particular cycle, as they had in every other cycle; so that we must then trace three distinct origins for the three distinct entities which unite to form man. One of these origins we have spoken of already, namely the moon. We pointed out that the moon was a living planet which passed through its life cycle and finally died, reembodying as the Earth, and that the hosts of entities that now build and inhabit this Earth once built and inhabited the moon, and other spheres before that, long since vanished. Man, being one of these races of entities, then, traces his lineage in part to the moon. But hold — what was man doing during that long interim between these two embodiments of the planetary being? At the time of the death of the moon, all that it had given him was surrendered to the dying planet, yet there was something that carried him over, something that evidently owed no allegiance to the moon; something that was imprisoned in the body of the then moon-inhabitant, just as there is something imprisoned within our bodies and personalities today, and which is liberated at death, and carried over for future lives. This something we may call the "soul," or better perhaps, the Higher Self. This Higher Self owes allegiance to none other than that bright cosmic entity focusing its energies through the sun, of whom we have spoken in earlier chapters. Thus, taking man as a whole, he owes his origin in part to the moon and Earth, but in a higher portion, to the sun.

Let us take once more our analogy of the reflected sun in the clear pond. When the pond dries, what has become of the reflected sun? It is no less a part of the solar energy, because the sun still shines. That solar energy always pervades the whole solar system. Let us say that the pond represents physical man. The water is part of the material of the Earth, yet the reflection is part of the energy of the sun. The drying up of the pond represents the fading out of the energies of the one-time moon. In like manner, at the death of the moon, man, as a race of entities, surrendered his lower portions to the source from which he drew them, and his Higher Self likewise withdrew to its source, entering into the very fabric of that bright cosmic energy, a part of whose aura we recognize as our sun.

Then, there is the third entity, what we call the Inner God, whose home is the spaces of Space, and whose origin is a star — its parent star — in our own Milky

Way perhaps, or maybe in a realm that is at present invisible to us. Verily, as the mystics say, so many men on earth, so many gods in heaven.

We are to think, then, of man as representing three distinct lines of evolution, each part of him being an evolving entity, and all linked by karmic bonds forged aeons ago. His destiny is to bring to comparative perfection this thing of earth and heaven we call man, which is only another way of saying to bring to us a full knowledge of THAT which is within and around and behind us all the time.

Now what happens in death is only a brief summary of what took place in those stirring times when the earth was young. A man dies; he throws his body aside, returning it to the earth whence it came. That lowest portion of the triad (which includes his unconquered passions, desires, and emotions) returns to those realms it recognizes as its home, leaving the higher self to enter into the very structure of the Cosmic Being from whence it came, and which it feels is its home. The inner God is drawn into the essence of its own home, the star of which it is a ray, and which indeed it never really left. Thus, in the space of a few moments is enacted a drama which actually took hundreds of millions of years to be carried out on a larger scale when the moon died.

When humanity came in due course out of its deep sleep into which it had fallen at the death of the moon, and entered the arena of the earth's life, the lower third of the trinity awakened to active consciousness first, and set about the task of building for humanity the kind of body that would be fitting for the uses to which the higher self was destined to put it on earth. Men of that time were thus only the awakened moon-children, learning after many trials to make a body that would be suited to the needs of man in the future. They might be called humans only by courtesy, for as yet they had not embodied the human faculties of intellect and reason. Their consciousness was childlike. In passing through these early stages, their forms underwent many changes, many of them resembling animals no doubt (hence the extreme difficulty of the problems before the anthropologists and the archeologists); but never were they animals in the full sense of the word. They were humans trying to express their humanhood; while the animals are animals first and last.

Just as the body of a child is too young to fully enshrine the intellectual and spiritual powers of the fully grown person, so the infant humanity of the earth-childhood was not evolved enough to enshrine the higher self. Even now this higher self is only just beginning to assert itself in the great masses of people.

Thus it took a long time for the higher selves of the human race to incarnate and make their presence felt. To be sure, they watched over infant humanity, but could not take an active part until the lower portions of themselves, the entities from the moon, had sufficiently evolved to allow them to enter as part of the human constitution. Even yet, man is but very imperfectly incarnated. He is too often a mere bundle of desires. The Master of Life is one whose higher self has fully incarnated, while the Super-Adept is one whose Inner God is likewise incarnated, so that in such a one you have a complete union of all the faculties, physical, mental, moral, spiritual, and divine. Such a one is a demigod like those of whom the myths and legends speak.

This slow process of evolution brought man to a certain point on each of the globes during the first Round, to a point a little more advanced during the second Round, to a point still more advanced during the third Round, and to the present point of advancement during this, our fourth Round. The possibilities of development can only be guessed, and the fifth, sixth, and seventh Rounds promise to bring forth a humanity that, towards the end of its career at any rate, will know its kinship with the gods. That, however, depends upon man himself and on whether or not he *wants* to know this divine kinship.

One of the main reasons for the promulgation of the Theosophical doctrines is to teach man to see before it is too late that he has a destiny to work out, and that if he fails, he will be left in the rear; for nature will not be held back on man's account. Oh, what a work there is to do before humanity even cares about these things! What a price we pay for ignorance and indifference! Just now, when the opportunity is at hand, we close our eyes to it, and trust to "God" or some outside power to take care of us to the end.

Before leaving this chapter, it might be well to point out the reason for limiting ourselves to considering the Earth Chain as though it were composed of seven globes instead of twelve. We held that these seven may be called the manifested globes in contrast to the five higher globes, which we call the unmanifested. A better explanation might now be made in the light of what we have just studied with regard to the three lines of evolution functioning within man, which we call, in the language of St. Paul when he echoed the ancient teachings, Body, Soul, and Spirit. You will remember that we have stressed a number of times the fact that what takes place in man is a copy of what takes place in the universe. A principle of life that operates for one, operates for all. In this manner, we may speak of the four lower cosmic planes of the Earth's constitution — where the seven globes before referred to are functioning as comprising the "body" of the earth ("body" in this case meaning the astral-vital-physical portion of its constitution). The two higher planes immediately "above," in which are functioning four of the higher unmanifested globes, may be called "soul." This will explain why it was better not to include them in the description of the manner in which the seven classes of entities form the Earth Chain. The "spirit" of the Earth Chain we have not as yet touched upon, and about it almost nothing can be said, except perhaps to speak of it as the focus of all of the highest energies of the Cosmic Being embodying as the Earth Chain. Diagrammatically, we represent it as the highest of all the twelve globes, thus:

O — SPIRIT

O O
O O — SOUL

A O O G
B O O F
C O O E
D O — BODY

It might be well, for the sake of review, to place here the paradigm used in Chapter V, showing the division of man's constitution into the triad, body, soul, and spirit.

Atman	Divinity	Spirit
Buddhi	Spiritual Soul	Soul
Manas	Mind	
Kama	Emotions, etc.	Body
Prana	Vitality	
Linga-sarira	Astral pattern	
Sthula-sarira	Physical Body	

XIV

Man, an Image of the Divine

The reader will recollect that we have studied man as being a threefold entity, *body, soul,* and *spirit*. This we found to be in reality a septenate as follows:

Atman	Divine Self	Spirit, the Inner God
Buddhi	Spiritual Soul	Soul, the Intermediate
Manas	The Thinker	nature of man
Kama	Desires, Emotion	
Prana	Vitality	Body, the substantial
Linga-sarira	Astral Pattern	portion of man
Sthula-sarira	Physical Body, built upon the Astral Pattern	

Each of these human principles has a cosmic counterpart, or we might say that each is the counterpart of a cosmic principle or plane of consciousness. To begin with, spirit, soul, and body correspond to gods, monads, and atoms. Each of these names refers to a certain class of beings. The gods are bright cosmic entities, long since evolved out of the human stage, and filling their appointed destinies, seen and unseen. The monads are the intermediate beings in the universe, neither the highest nor the lowest. There are innumerable classes of these monads, one instance of them being those who ensoul our humanity here on earth. The term "atoms," refers to those countless lives which compose the vehicular portion of the universe; they are the substances of which are made the bodies — whatever kinds they may be — of the highest beings which embody themselves whenever and wherever their destiny calls them to do so. There are definite correspondences between the functions of these three main streams of evolution in the universe and the three main lines of evolution in every human.

One point that we shall endeavor to explain here is that these human principles are not to be considered as entirely separated from one another like layers in a cake, but they merge and blend into one another at the various centers of consciousness that we are now to consider, and which we call "souls" and "egos." These souls and egos are really the workings of these principles as their energies unite one with another.

Now Theosophists use the world "soul" in a very technical sense. They do not mean by the word merely the indwelling consciousness within a person which survives after the death of the body. They mean a vestment, a vehicle, of the spirit within. And, to explain further, each plane of consciousness within the person is dual. Each has its own energic and its own substantial aspect.

The reader will recollect that in one of the earlier chapters of this book we learned that every plane of consciousness in the universe is bipolar, or composed of spirit-matter of various grades according to the plane in question. Later on we expanded this teaching, telling how the one aspect of each cosmic plane in turn served as a portion of the "downward" or "shadowy" arc for the hosts of life-energies pursuing their way through the globe chain, and the other aspect of these same cosmic planes served as a portion of the "upward" or "luminous" arc for those same entities on their upward journey through the higher realms whence they had originally sprung. The study of these planes of consciousness in their dual aspect is technical indeed, and will be only lightly touched upon in this book, but it brings in what are termed the "lokas" and "talas," the lokas being the energy-aspects of the planes, and the talas their substantial aspects. In the constitution of our Earth Chain taken as a whole, the talas are symbolized by, or rather their attributes are focused in, Globes A, B, C (and to an extent D), forming the "downward arc," and in the globes of the upward arc, E, F, and G, we have the focus of the energies forming the lokas, or spirit aspect of the four lower planes of consciousness.

In man we also have the lokas and talas. This means that every one of his little planes of consciousness, his principles, is likewise dual; the energy aspects of these human planes of consciousness we call the "egos," and the substantial aspects we call the "souls." With these facts in mind, we will now revise once more the tabulation of man's consciousness aspects as follows:

Atman	The Self		Spirit
Atma-Buddhi	Spiritual Ego	Spiritual Soul	
Buddhi-Manas	Human Ego	Human Soul	Soul
Manas-Kama	Personal Ego	Personal Soul	
Kama-Prana	Animal Ego	Animal Soul	
Prana-Linga-sarira-Sthula-sarira	Vital-Astral-Physical Body		Body

We must now tabulate, as well as can be done, the

seven planes of consciousness forming the universal constitution:

Brahman	The Cosmic Self, root of all that composes the universe.
Maha-Buddhi	The source of all the "laws" of the universe. The intelligence pervading the universe.
Mahat	Universal Mind. The source of all monadic individualities in the universe.
Fohat	Cosmic Kama, the vital desire to become, which brought about the growth of the universe.
Jiva	Cosmic psycho-electric-magnetic field, from which all individuals derive their respective pranas.
Astral Plane or Astral Light	The universal pattern on which is built the grossest or physical universe. The astral light performs many functions, among which is the storing of all the efflux from the psychological, mental, emotional, and even physical planes.
Physical Plane	The plane of objective life. The outermost garment of universal being. Actually, concreted astral substance.

In order to bring these correspondences more forcibly to mind, we shall make use of the lens diagram previously employed, by means of which we attempted to show that the principles of man's constitution are derived from the cosmic principles or elements to which they correspond, indeed with which they are fundamentally one. We may now expand the diagram used in Chapter VI in which we thought of the essential consciousness of man as though it were a lens with the power to focus the cosmic energies into an image of the universe, much as a burning glass focuses the rays of the sun, forming an image of the sun, which is in effect formed of the same energy as the sun itself. This "image of the universe" is man himself. Thus in our new analogy diagram, the universal consciousness, Parabrahman (meaning that which is beyond Brahman, or the causeless Cause of all that is), is brought to a focus, and the image is Paramatman, the root of all that man is. Paramatman bears the same relation to the whole constitution of man as does Parabrahman to the whole range and gamut of planes of consciousness composing the universe.

Now an image of the sun, such as is cast by a lens, is not a mere fantasy, a mere illusion. It has reality, and is actually a focus of energy, evidenced by the fact that it can be used to ignite wood or paper. The rays of light composing it are none other than the solar rays passing through the lens. Likewise, Paramatman, this causeless Cause of man's nature and evolution, is actually a part of the universal root of all, Parabrahman. And just as the light from the sun itself may be separated into seven prismatic colors, so may the light of the image of the sun be broken up into the seven colors, because it is all part of the one solar energy. In our analogy, the seven energies emanating from Parabrahman are found to be the source of the seven

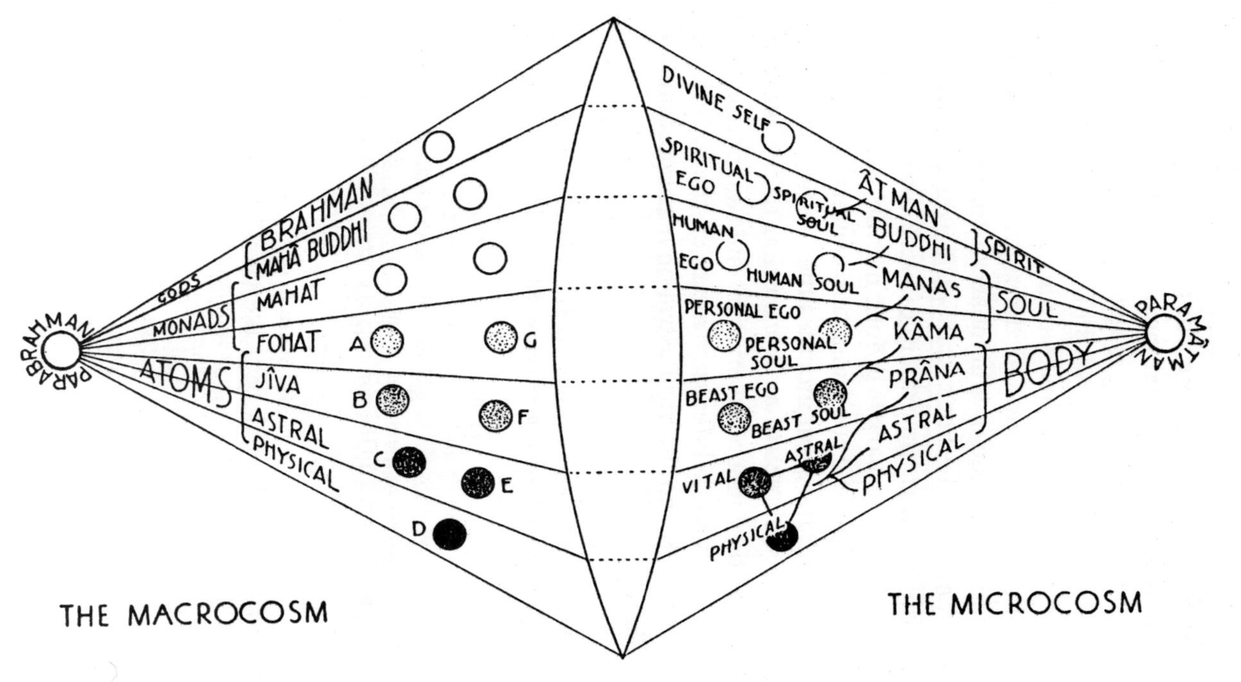

PARABRAHMAN
GODS
MONADS
ATOMS
BRAHMAN
MAHÂ BUDDHI
MAHAT
FOHAT
JÎVA
ASTRAL
PHYSICAL
A
B
C
D
E
F
G
DIVINE SELF
SPIRITUAL EGO
SPIRITUAL SOUL
HUMAN EGO
HUMAN SOUL
PERSONAL EGO
PERSONAL SOUL
BEAST EGO
BEAST SOUL
VITAL
ASTRAL
PHYSICAL
ÂTMAN
BUDDHI
MANAS
KÂMA
PRÂNA
ASTRAL
PHYSICAL
SPIRIT
SOUL
BODY
PARAMÂTMAN
THE MACROCOSM
THE MICROCOSM

energies emanating from Paramatman, because they are all one and the same thing in essence. We then complete our diagram, filling in the various items we found in the table above.

There are a few points that should be mentioned. In the first place, looking at the right-hand side of the diagram, the reader will observe that the brackets designating Body, Soul, and Spirit overlap; further, it is indicated that the principles work together, two by two for the most part, in forming the souls and egos that are really the consciousness centers in man. Thus we see that the *spirit* includes the divine Self, and the spiritual soul, spiritual ego portion of the man; *soul* includes the human ego, human soul, and personal ego, personal soul; while *body* includes the animal ego, animal soul and the vital-astral-physical portion of his makeup.*

---

*It may be of interest to some to observe that *body* includes four principles, *soul* includes three, and *spirit* includes two. These digits, 4, 3, 2, are master keys in computing the occult numbers relating to the cosmic cycles such as are in the zodiac; earth cycles, such as are in the Rounds and races as the monads pursue their evolutionary journey through the globes of the Earth Chain; human cycles such as the breathing and heartbeat; and to the cycles in the still life of crystals, wherein the numbers of degrees in the various angles in the regular polyhedra are seen to be built upon the number 432. Instances follow: cosmically, there are 2,160 years in the passage of the equinoctial point through one sign of the zodiac. 2,160 is ½ of 4,320. The life of the Earth is 4,320,000,000 years (called in the old Brahmanical reckoning the "Day of Brahma"). The human heart beats on an average 4,320 times in one hour, and the breathing count is 18 times a minutes, which makes 4,320 times in 4 hours, and 25,920 times in one day. (25,920 is the number of years in the complete zodiacal cycle, in which the equinoctial point on the Earth's orbit will have passed entirely around through the 12 constellations of the zodiac.) And lastly, in solid geometry, even in the cube, we have 2,160 degrees as the sum of the plane angles — ½ of 4,320. These are but a few instances that might be cited to show the interrelations of cosmic forces.

What we have given here must of necessity serve as a mere hint or suggestion of the value of correspondences, it being left to students to turn to the more recondite teachings if they have the real desire to follow these things further. The Ancient Wisdom is to be had for the asking, but one must know how to ask.

XV

The Powers Innate in Man

In taking the threefold division of man's nature as a basis upon which to build, expanding it into the sevenfold (as in Chapter V) and then into the twelvefold (as in Chapter XIV), we have made an effort to demonstrate the fact that man is essentially a child of the universe, drawing his life and substance from the three main evolutionary streams, to wit, gods, monads, and atoms. Thus, as we have seen, a man represents three main lines of evolutionary development, spirit, soul, and body — using the terms in their broadest sense.

We have, moreover, endeavored to set forth the teachings concerning the origin of these three portions of man's constitution, explaining that *spirit,* the godlike portion of man, is of the same essence as the gods themselves, and is therefore really the Inner God of the man. Verily, "It takes a God to make a man." This Inner God is in reality a ray of a star — not a light ray, of course, but a spiritual-divine ray of the spiritual-divine consciousness working through a star. Around this fact is woven the teaching of the "Parent Star," a matter to which the Teachers merely allude, as it pertains to the most recondite teachings of the Mystery

Schools, and has therefore been carefully guarded by occultists in all ages. And who are they to tell their secrets?

Soul, we learned, is of solar substance, belonging as it does more specifically to our solar system with which it is karmically linked throughout the entire life cycle of the solar system. It is this fact which will explain why we are drawn back to earth at the time of reincarnation, for that part of us which reembodies never really leaves the solar system.

Body is that portion which includes the vital-astral-physical streams of evolution, and which in itself is evolving, because it too enshrines a consciousness — a more elemental consciousness, to be sure, but one nevertheless with a destiny before it, so that it too will some day function as the *soul,* and again, in the still more remote future, as the *spirit* of a future human entity, evolved from its present self, which will play its part on the cosmic stage aeons hence.

Up to the present, however, we have not touched upon the powers of the various portions of man. Once having done so, we can the more readily understand the process attending the change called death, and outline to some extent the destiny of man, as the natural result of self-study and self-directed evolution. That is our present purpose, and the study is fascinating indeed. And why should it not be so? We are men and women taking no small part in the cosmic drama going on about us, rehearsing nightly in sleep the role we are to play in death, rehearsing in birth the part we are to play in the building of new worlds, ages hence.

Briefly, then, the powers in man may be set forth as follows. Let us begin with those of the *body.* There are few powers originating in the body itself, as it is mainly "the playground of the senses." These senses are not in themselves physical, but are more strictly astral, and function through the appropriate physical organs. Indeed, we can hardly consider the physical body as an individual human "principle," for it is in reality composed of the lowest grades of astral substance. Physical matter is to be distinguished from astral matter only in degree of grossness. Actually the two are so much one and the same thing that whether consciously or not we are living on the two planes at the same time. This fact would be readily understood had we more sensitive organs of perception. The powers that belong to the physical body as such are really limited to those which go toward building and maintaining the structure and health of the body, yet even these powers have their origin in prana, or the vitality. We may draw a lesson right here, for we find it difficult to separate the forces and powers of man's constitution into little airtight compartments, as it were. It simply cannot be done. The best that can be accomplished along this line is to show how the one life operates through the entire system, manifesting under various forms in the various departments of man's nature.

The powers of the astral body are largely automatic in the spiritually undeveloped man, although in the future (for the time has not yet come) men and women through suitable methods of training can learn how to utilize the astral body, making of it a new instrument by means of which they may gain higher knowledge than is possible while chained as they are today in the physical body alone. At the present time, however, the astral body is the pattern around which is woven, cell for cell, the physical body. During the prenatal period of growth, the vital forces join with the astral to form the physical, and in some respects the vital powers are in their heyday at that time. Of course they have their work to do through the whole life of the man, but some of the vital currents sink into relative quiescence after birth. Whereas they have acted automatically previous to that time and therefore function in full accord with nature's plan, with the result that a healthy and well-formed body is prepared for the incoming man, one can readily see how dangerous it is to meddle psychically with these pranic powers, and try to awaken them into a forced activity during the normal life of a person. There are most powerful vital forces locked up within us all, and in the natural course they should be allowed to function in their normal ways. There are certain methods whereby they may be awakened and the "vital airs," as they are called, be caused to flow more quickly, but exponents of these quasioccult powers almost invariably suffer in the end, for, acting out of time, these powers destroy instead of build, and the result is disease of both mind and body.

Then there are the intense desires and emotions ready to leap forward at a moment's notice, which wreak havoc in the world today, and yet they are to be numbered among the powers of man. Thus far we have enumerated the powers of the lower portion of man's constitution, that part we have referred to as "body."

It may be argued by some who have developed their psychic powers to some extent that wonderful things can be accomplished. Yes; they may be able to create a sensation among the ignorant, but they should learn that their powers are limited to a very small sphere of consciousness, and a low one at that. Moreover they are leading others as well as themselves into dangers that they know not of. Never by the use of their awakened "vital airs," or psychic senses, will they be able to rise to the lofty planes of being on which are operating the forces that keep mankind from destruction. However clairvoyant or clairaudient they may have become on the physical or on the astral plane, they will never catch the least strain of the "music of the spheres," nor even a glimpse of the armies of monads as they pursue their journey through the spaces of Space. These things are open to the one who has developed the powers residing in the soul and spirit portions of man, and which we shall now consider.

Soul, as you will recollect, consists in part of the higher mind illumined by the light of the Spiritual Soul. This union is what eventually makes a Master of Life. Thoughts are things, elemental consciousness-entities, alive and evolving, and therefore are a real power in the world. The higher mind is capable of entertaining sublime visitors in the shape of lofty ideas which have the power to raise the being to the heights of spiritual understanding. When the power of right thinking has been developed, further steps follow. They lead the student to a new understanding born of experience. For, just as there is a kind of clairvoyance on the astral plane, so there is a type of true clairvoyance on the spiritual and cosmic planes, but the way to the development of this true clairvoyance is a long road of effort and altruism, and the things that may be discerned by those who have developed these spiritual powers of perception become lessons of life. They hear the symphony of life eternal.

One of the less important phases of the spiritual power of inner perception is one's ability to communicate over any distance whatever with others who have developed the same powers. Moreover, one may travel to any point on the earth at a moment's notice, by using the highly developed astral body before alluded to, and which is more technically called the mayavi-rupa, because it is more than mere astral substance. It is fashioned of mind-stuff as well. In this mayavi-rupa, we are told, a person is able to carry on work whenever and wherever he is needed, and he may be seen or not as he chooses, although he probably would appear objectively to those only who are engaged in the same lofty work as he. This work we shall treat in our next and final chapter.

The powers of the Inner God of man are mighty indeed, and by their aid the Adept is able to penetrate into the inner and invisible realms of nature, and consciously to communicate with those divine beings we call the gods. What a sublime experience that must be! To know what the gods are thinking, because you think like them! To know the work they are doing, because you participate in that work! Is this not a sublime adventure?

XVI

The Masters of Wisdom and Their Work

Tempting as it might be to trace the history of mankind back through the ages and recount the achievements of past civilizations, such a task is beyond the scope of this present book. However, we can and will cull a few fragments of forgotten truth concerning man's spiritual progress, and bring them to bear upon the problem in hand, to wit, the setting forth of ideas totally strange to many in the West.

It should be remembered that history is more than a mere record of events and dates, however important these may be. There is an occult history of the human race which is the record of the gradual unfolding of man's inner powers, which in their turn resulted in the doings that are set forth in the history books. Some day an occult chronicle may be written which will divulge some little-known facts about the early races of man. Suffice it to say now that the whole scheme of the history of our earth and its races as framed by archeologists and geologists today is built upon a wrong premise. Evidences have been found that caves were once occupied by people of a prehistoric time, and the conclusion formed is that man had his humble beginnings in company with the animals from which he had sprung, and that in the course of a few millions of years he developed his present-day intelligence. The skulls that have been found in various parts of the world seem to tell their story of the gradual change of the shape of the head and general stature through the ages, and these findings are set forth as though they were fully established facts.

How different is this story from that related by occult science, which points to a period in world history predating anything that has been even remotely suspected by modern science. It declares that nothing has been found in the way of human remains from those remotest times because the humans of those times left no remains. Their bodies, like the earth itself, were composed of a much finer grade of astral substance, so that whatever records were left remain in the astral light or the astral sphere surrounding and interpenetrating the gross physical earth. There these records are, and they may be seen today by those who have had the training to see them. But they will never be seen or understood by those who laugh irreverently at truth when that truth does not fit their notions — notions built upon the insecure rock of so-called proof. If you wish to prove a thing to another, you may or may not be able to do it, for you do not know another's mind. What is proof for you is not necessarily proof for another. But if you wish to prove a thing for yourself, you have at least to keep an open mind and pursue your studies in accordance with the methods which are necessary to the subject which you are studying.

Thus the occult science points to civilizations of the past which had their flowering and their seeding time, just as ours is approaching its flowering, and will in turn pass through its time of seeding. Each new civilization springs from the seed of the old. Old civilizations are not therefore utterly lost, for, due to the law of reincarnation, we, the members of the present civilization, were members of the civilizations of the past.

At the present time we are passing through a particularly trying cycle, a time of transition really, and although these truths arc difficult to receive, yet there is an opportunity to learn such as we have not had for centuries. The wisdom of ages is our heritage, and woe to the human race if we do not claim what is rightly ours and draw upon the treasury of wisdom contained in the ancient mysteries, and shape our lives in accordance therewith. Here are the facts then, as we have been taught them.

There was a time in human history ages ago when materialism had not yet set in, when an instinctual sense of the mysteries of nature was the common property of all. Intercourse between the gods and humans was easy, and many were aware of their divinity. As time went on and developing man descended farther into matter, and the thickening veils of selfhood obscured the face of the divine, this knowledge would have been lost forever had it not been that a brotherhood was formed of the more advanced individuals of the human race for the purpose of preserving the wisdom of the gods. This band of holy ones, whose lives were given to the cause of saving the wisdom handed down from the gods for humanity's future use, carried on its work through the rise and fall of nations and of civilizations.

In each generation these mighty ones train their disciples so that the band shall not die out, teaching them how to guard the secrets. They search at all times for unselfish souls who are willing to give themselves and subordinate their destinies to the welfare of humanity, ever and anon giving out such teachings as mankind is ready to receive. Each age has brought forth a few such self-sacrificing ones, so that this brotherhood has never died out. History is brightened by the appearance of great Teachers, such as Gautama the Buddha, Jesus, Confucius, Lao-Tse, Pythagoras, Apollonius of Tyana, and many more. These had a common work. Not only were they in search of new recruits, but they came among humanity to set in motion new currents of thought, to help people raise their minds and hearts to new heights of understanding and sympathy.

Study the great religions of the world, and also the great philosophies, and you will find that, framed as they were in different manners to suit the times, the purport of their message is that people should learn to *know themselves,* know life, know and understand nature, and, knowing, learn to live. Furthermore, these had definite technical teachings to offer to supplement the ethical doctrines, for goodness must be enlightened by knowledge if the redemption of humankind is to become a reality.

And what does all this mean to us of the present day? It means this: Just as in the past there was need of new recruits for the work, so there is that need today, and there is no one who is not in a position to help in some measure. If we have any light at all — and who has not? — it is our bounden duty to share that light with those who have less than we. And the reward? The inexpressible reward that comes is the growing ability to teach, to exemplify, to instruct in the Divine Knowledge.

The student is assured in all solemnity that there live great ones today, those of wonderful caliber — supermen if you like — whose secret home is in the fastness of the Himalayan mountains, in a land called

Shambhala, of which there are many legends. Not only is their home there, but they and their disciples may be found all over the world, wherever they are carrying on their work. There are more of these than one might at first imagine. And what is their influence in the world? They help wherever their help can be received. Wherever there is genuinely humanitarian work going on, whether along religious, philosophic, scientific, or other lines, their influence is felt, sometimes consciously, though more often unconsciously or only semiconsciously, by those who have the high privilege to be in a position to work constructively for the good of humankind.

The work done by these Great Ones is of varying types. Some of it is of a general nature, as has just been explained. A certain few have a particular work to do in the world, and a notable example of this is to be found in the lives of two of these great men who gave the originating impulse to the present Theosophical Movement, and who today vitalize it with their own spiritual and intellectual energy. These two are those who are known among Theosophists by their initials M. and K. H.

Then there is a still more particular work that these great ones are doing in their spiritual home, Shambhala. But of this only hints have been given to us. We are told that they form a guardian wall of human hearts, protecting the earth against cosmic influences which, if allowed to play freely upon the earth, would soon destroy humanity and indeed all life. Yes, many conditions are discouraging today, but they are nothing compared to what they would be had we not the protecting influence of the guardians who are ever on the watch.

Now, these things may be difficult to accept unless we are assured that we can know of them by personal observation and experience. And there does not live the man or woman who cannot ultimately know these things for himself or herself. Obviously there are the correct means to an approach to the sanctum. The approach is paved with deeds of self-forgetfulness and a will to know that cannot be turned aside. The selfish plea for knowledge will never be heard. These things must be sought after for the one purpose of using the knowledge thus gained for the betterment of the human race, and that the heavy weight of suffering and sorrow may be lifted.

Then comes the laudable desire to know more about the Adepts, or Mahatmas, as they are called. There is only one way to know them. Find out what their work is, and take part in it, and thus you will become like them, and approach nearer to them. Seeing an Adept will mean nothing unless you have the power to recognize him as such. "It takes greatness to recognize greatness." Were an Adept to appear on the streets of a crowded city, we might never recognize him at all unless we knew how to look deeper than the outer appearance. And we need not think that they all have long beards and mysterious eyes. If any are working in the Western world at present, they probably are doing so as members of our Western civilization, and very unpretentiously, for what they least desire is to create a sensation. We have but to join forces with them in their sublime work, and, learning how to raise our consciousness to the plane on which they act, we will discover how to recognize them. Recognition of the Adepts by anyone means that within such a person is something that is akin to the nature of the Adept, and this is only another way of saying that he or she is an Adept in the making. And who may not become such?

Now there are schools here in the West, as well as in the East, where definite knowledge of these things is to be had. The teachings are never to be paid for, except in devotion to the cause of universal brotherhood. But it requires an iron will and a self-forgetful heart to carry the questing pilgrim to the goal. Anyone may make the start. Anyone at all. But how many will reach the end? We are assured that in these Mystery Schools of today no psychism or phenomenal magic is taught, no postures, or similar methods of concentration. But there is the white magic of spiritual thought, and the practice of ethics first and last. Also technical teachings are given about the spiritual nature of man and the universe, and the training of the mind and heart to make these teachings a living part of the life of the student. When all of these conditions have been fulfilled, and beginners have learned how to pledge their whole being to their own Higher Self and to the Masters' Cause, then and then only may they hope for further progress. The Great Ones are ever looking for those among our race for whom the ideal of human perfection is the North Star by which we set our course.

Index

QUEST BOOKS
are published by
The Theosophical Society in America,
Wheaton, Illinois 60189-0270,
a branch of a world organization
dedicated to the promotion of brotherhood and
the encouragement of the study of religion,
philosophy, and science, to the end that man may
better understand himself and his place in
the universe. The Society stands for complete
freedom of individual search and belief.
In the Classics Series well-known
theosophical works are made
available in popular editions.